# Loch Leven: 40 Years of Scientific Research

Developments in Hydrobiology 218

*Series editor*

K. Martens

For other titles published in this series, go to
www.springer.com/series/5842

# Loch Leven: 40 Years of Scientific Research

*Editors*

Linda May & Bryan M. Spears

*Centre for Ecology and Hydrology, Penicuik, United Kingdom*

Previously published in *Hydrobiologia*, Volume 681, 2012

Springer

*Editors*
Dr. Linda May
Centre for Ecology and Hydrology
Edinburgh Bush Estate
EH26 0QB Penicuik Midlothian
United Kingdom
Email: lmay@ceh.ac.uk

Dr. Bryan M. Spears
Centre for Ecology and Hydrology
Edinburgh Bush Estate
EH26 0QB Penicuik Midlothian
United Kingdom
Email: Spear@ceh.ac.uk

ISBN 978-94-007-4332-8 ISBN 978-94-007-4333-5 (eBook)

Springer Dordrecht Heidelberg New York London

Library of Congress Control Number: 2012934454



Printed on acid-free paper.

Springer is part of Springer Science+Business Media (www.springer.com)

# Acknowledgements

The guest editors are grateful to the following reviewers for their help in preparing the manuscripts:

Alistair Dawson, Christian Feld, Glen George, Chris Goldspink, Ramesh Gulati, Liz Haworth, Harry Hosper, Ken Irvine, Andrea Kelly, Alan Lauder, Suzanne McGowan, Stephen Maberly, Don Monteith, Geoff Phillips, Martin Søndergaard, István Tátrai, Max Wade, Allan Watt and Nigel Willby.

We would also like to thank those reviewers who preferred to remain anonymous.

# Contents

Hydrobiologia (2012) 681:1–2
DOI 10.1007/s10750-011-0928-5



# In memoriam

## Antony Edward Bailey-Watts PhD, D.Sc. (23 August 1942 – 15 July 2007)

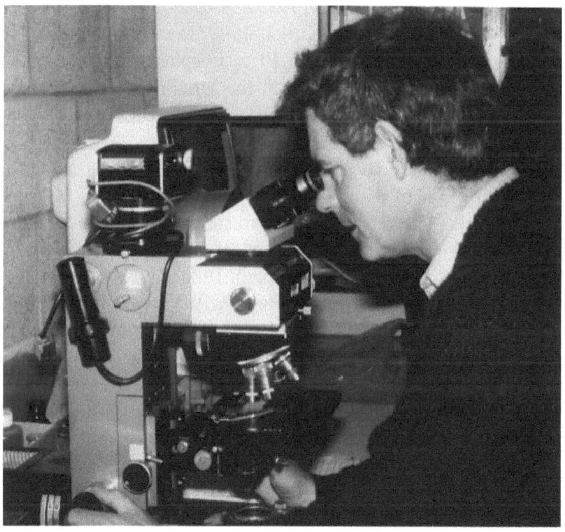

This volume is dedicated to the memory of Dr. Tony Bailey-Watts who died on 15 July 2007. The papers that follow would not have been possible without Tony's vision, his commitment to long-term limnological research at Loch Leven, and his extensive legacy of scientific publications.

Tony began work as an algal ecologist at Loch Leven on 1st April 1967, having graduated from Sir John Cass College (now London Metropolitan University) and following a brief period of work for the River Dove Water Board, UK. His contract was initially for a 1 year period, working on Loch Leven as part of the International Biological Programme (IBP). IBP was one of the first collaborative projects designed to investigate shallow lake ecology across the world. Tony was offered a permanent research post in 1973, the same year that he was awarded a PhD for his work on Loch Leven by the University of London.

These early years were extremely productive for Tony. He published his first paper in the journal Nature (Bailey-Watts et al., 1968) and met colleagues and formed friendships that were to last throughout his life. Tony's training during this period could not have been more illustrious with names in the acknowledgements section of his PhD thesis including Margaret Bindloss, Peter Maitland, Hugh Cary Gilson, Jack Talling, Frank Round, Ian Smith, Hilda Canter and John Lund. John Lund, in particular, had a strong influence (both academically and socially) on Tony throughout his research career. His research was also strongly supported and encouraged by Sir David Montgomery, the owner of Loch Leven, and Alex Kirika, who conducted the routine sampling and analyses at the site for more than 30 years before retiring in 2007.

During his 33 years of working on Loch Leven, Tony published 70 scientific papers, around 30 of which were focused on the ecology of the loch. This body of work encompassed a wide range of disciplines

Guest editors: L. May & B. M. Spears / Loch Leven: 40 years of scientific research including eutrophication and eutrophication management (Bailey-Watts, 1990, 1994), fisheries management (Bailey-Watts & Maitland, 1984; Bailey-Watts et al., 1987), biogeochemical cycling between the sediments and water-column (Bailey-Watts & Kirika, 1996), metal contamination of freshwaters (Bryant et al., 1991), computer aided collection and management of algal counting data (Robson & Bailey-Watts, 1987), catchment nutrient loading processes (Bailey-Watts & Kirika, 1999), phytoplankton ecology (Bailey-Watts, 1997, Bailey-Watts, 1998), food web structure and function (May et al., 2001), paleaolimnology (Eades et al., 2002) and socioeconomics (Bailey-Watts 1990). In addition, Tony published on a new species of the cyanobacterial genus *Synechococcus* (Bailey-Watts & Komarek, 1991). This incredible scientific legacy continues to drive research on freshwater shallow lakes, today, and is still used as a template for the management of these complex systems.

The IBP monitoring work clearly had a major influence on Tony's research and his view on the importance of long-term monitoring in providing the data with which scientific ideas could be validated is reflected in many of his papers. It is, however, this strong dedication to long-term monitoring at Loch Leven, especially during periods when funding bodies did not share those views, which has resulted in the internationally important dataset that we have today. This enthusiasm and dedication to the cause has been passed on to many of Tony's students and colleagues over the years, most notably during his work for the Lake Tanganyika Biodiversity Project (1995–2000) and through his involvement in the International Association of Phytoplankton Ecology and Taxonomy.

Tony's outstanding contribution to the field of limnology continues to inspire, educate and shape research at Loch Leven and many other sites. Tony's influence and views on lake restoration still reverberate through the scientific, policy and business communities at Loch Leven and we hope that the following group of papers help to advance his work.

Finally, should any future students or researchers find themselves in need of inspiration, we recommend that they read one of Tony's papers.

Bryan Spears & Linda May
Centre for Ecology & Hydrology, Edinburgh, UK
e-mail: lmay@ceh.ac.uk

## References

Bailey-Watts, A. E., 1990. Eutrophication: assessment, research and management, with special reference to Scotland's freshwaters. Journal of the Institute of Water and Environmental Management 4: 285–294.

Bailey-Watts, A. E., 1994. Eutrophication. In Maitland, P. S., P. J. Boon & D. S. McLusky (eds), The fresh waters of Scotland: a national resource of international significance. Wiley, London: 385–411.

Bailey-Watts, A. E., 1997. The phytoplankton ecology of the larger Scottish lochs. Botanical Journal of Scotland 49: 397–404.

Bailey-Watts, A. E., 1998. The phytoplankton ecology of the larger Scottish lochs. Botanical Journal of Scotland 50: 63–92.

Bailey-Watts, A. E., M. E. Bindloss & J. H. Belcher, 1968. Freshwater primary production by a blue–green alga of bacterial size. Nature 220: 1344–1345.

Bailey-Watts, A. E. & A. Kirika, 1996. A brief review of Loch Leven studies and some new findings on the sediments as an unusual diffuse source of phosphorus. In Petchey, A., B. D'Arcy, & A. Frost (eds), Diffuse Pollution and Agriculture. SAC-FRPB Conference, University of Edinburgh, Edinburgh, 12–13 Apr 1995: 224–232.

Bailey-Watts, A. E. & A. Kirika, 1999. Poor water quality in Loch Leven (Scotland) in 1995, in spite of reduced phosphorus loadings since 1985: the influences of catchment management and inter-annual weather variation. Hydrobiologia 403: 135–151.

Bailey-Watts, A. E. & J. Komarek, 1991. Towards a formal description of a new species of *Synechococcus* (Cyanobacteria, Microcystaceae) from the freshwater picoplankton. Algological Studies 61: 5–19.

Bailey-Watts, A. E. & P. S. Maitland, 1984. Eutrophication and fisheries in Loch Leven. In Holden, A. V. (ed.), Proceedings of the 1984 Study Conference of the Institute of Fisheries Management, 11–13 Sept 1984 I. F. M., Stirling: 170–190.

Bailey-Watts, A. E., E. J. Wise & A. Kirika, 1987. An experiment in phytoplankton ecology and applied fishery management: effects on troublesome algal blooms of artificial aeration in a small eutrophic loch. Aquaculture and Fisheries Management 18: 259–271.

Bryant, C. L., J. G. Farmer, A. B. Mackenzie, A. E. Bailey-Watts & A. Kirika, 1991. The biogeochemistry of heavy metals in an acidified Scottish freshwater loch. In Proceedings of the 8th International Conference on Heavy Metals in the Environment, Ann Arbor 1: 347–350.

Eades, L. J., J. G. Farmer, A. B. Mackenzie, A. Kirika & A. E. Bailey-Watts, 2002. Stable lead isotopic characterisation of the historical record of environmental lead contamination in dated freshwater lake sediment cores from northern and central Scotland. Science of the Total Environment 292: 55–67.

May, L., A. E. Bailey-Watts & A. Kirika, 2001. The relationship between *Trichocerca pusilla* (Jennings), *Aulacoseira* spp. and water temperature in Loch Leven, U.K. Hydrobiologia 446(447): 29–34.

Robson, P. G. & A. E. Bailey-Watts, 1987. A computer–based system for recording and sorting phytoplankton count data. British Phycological Journal 22: 261–267.

Hydrobiologia (2012) 681:3–9
DOI 10.1007/s10750-011-0929-4

LOCH LEVEN RESEARCH

# A history of scientific research at Loch Leven, Kinross, Scotland

**L. May · B. M. Spears**

Published online: 4 November 2011


**Abstract** Loch Leven is a large, shallow lake in lowland Scotland, UK. Scientific research began here almost 200 years ago. Early research characterised the biodiversity and physical characteristics of the loch, providing an important historical background for future research. In the mid-1960s, this ad hoc approach was superseded by a more structured research programme under the umbrella of the International Biological Programme. This was the beginning of the Loch Leven long-term monitoring programme. Today, the results of these studies form one of the longest and most comprehensive limnological datasets for shallow freshwater lakes in the world, comprising more than 500 physical, chemical and biological variables collected at two-weekly intervals. To celebrate the 40th anniversary of the start of the long term monitoring programme, and to highlight the scientific investigations still being conducted at Loch Leven, the NERC Centre for Ecology & Hydrology (CEH) organised a symposium entitled "Loch Leven: 40 years of scientific research" in Kinross, Scotland, UK, on 11 December 2008. This examined the role of long-term monitoring in developing our understanding of the links between pollution, climate change and ecological responses in shallow lakes. This article introduces a series of papers summarising the scientific results presented at this meeting.

**Keywords** Long-term monitoring · Eutrophication · Recovery

## Introduction

Loch Leven is a large, shallow lake in the lowlands of Scotland, UK. There is a long history of scientific research at this site, which spans almost 200 years and has produced more than 140 publications (Fig. 1; Online Resource 1). This impressive body of work has advanced fundamental scientific theory on lake ecosystem functioning and provides a framework for the development of lake management systems.

Over the years, the main focus of research activities at this site has varied in response to the changing demands of the world in which we live. Early research was mainly focused on characterising the biodiversity and physical characteristics of this shallow productive lake, providing an important historical backdrop for future research activities. In the mid-1960s, this somewhat sporadic early work was superseded by a more structured and better focused research

Guest editors: L. May & B. M. Spears / Loch Leven: 40 years of scientific research

**Electronic supplementary material** The online version of this article (doi:10.1007/s10750-011-0929-4) contains supplementary material, which is available to authorized users.

L. May (✉) · B. M. Spears
Centre for Ecology & Hydrology, Penicuik EH26 0QB, Midlothian, UK
e-mail: lmay@ceh.ac.uk

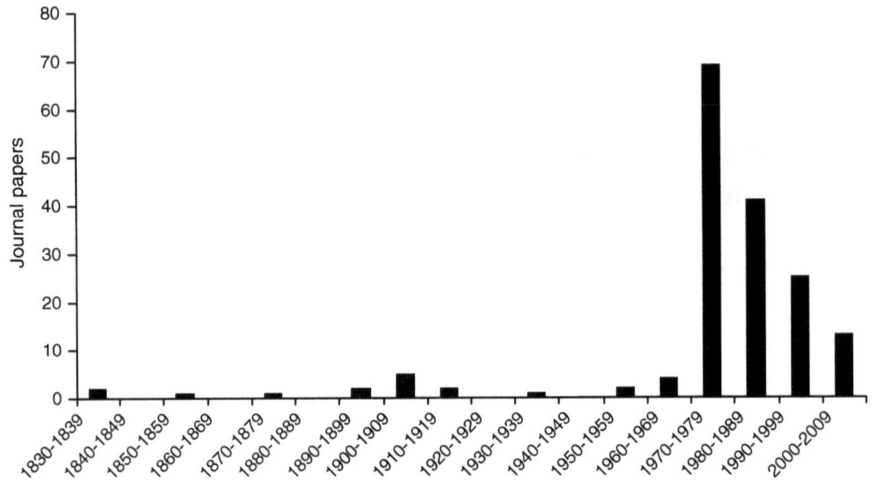


**Fig. 1** Scientific research at Loch Leven spanning almost 200 years has produced more than 140 publications

programme carried out under the umbrella of the International Biological Programme (Morgan, 1974). This was the start of what has now become the Loch Leven long-term monitoring programme. Today, the results of these studies form one of the longest and most comprehensive limnological datasets for shallow freshwater lakes in the world, comprising more than 500 physical, chemical and biological variables. This introduction provides an historical background to this programme, describing the scientific and conservation importance of the loch and giving context to the massive effort required to develop and, during darker times, simply maintain, this remarkable research programme.

## The conservation and socio-economic value of Loch Leven

Loch Leven has high conservation value at both the local and international level. In recognition of this, the site was made a site of special scientific interest (SSSI) in 1956 because of the large number of wildfowl that frequented the loch and the outstanding number of higher plant species and rare insects that it supported. In 1964, the loch was awarded the status of national nature reserve to "maintain its attraction for all species of wildfowl and to safeguard all the characteristics that go to form this unique habitat". In 1971, it was declared a Ramsar site, reflecting its importance as a naturally eutrophic loch and wetland system with international importance as an overwintering site for waterfowl, especially pink-footed geese (*Anser brachyrhynchus*) and shovelers (*Anas clypeata*). More recently, the loch became a Natura 2000 site, forming part of a network that aims to protect natural habitats and rare species.

In addition to its conservation importance, the loch has supported a world famous trout fishery for more than 150 years. Loch Leven trout were particularly attractive to anglers because of their dark colouration and unusually pink flesh and, at one stage, were classified as a distinct species, *Salmo levenensis* (Yarrell, 1839), although this is no longer case. Loch Leven trout became world famous for their quality in the late 1800s and early 1900s and, as such, were exported to fisheries around the world. There are many historical records of these fish being introduced into lakes in north and south America, many Asian countries and as far south as New Zealand. At that time, such translocations were seen as positive, newsworthy, events, as demonstrated by this report which was featured in the New York Times on 23 December 1884:

> "The Anchor Line steamer Furnessia, which sailed from Glasgow yesterday for New York, has on board 100,000 Loch Leven trout ova, which are a present from Sir Gibson Maitland to the American Fisheries Commission, to be hatched in Michigan and introduced into the Great Lakes of America".

The historical significance of these activities are proving to be just as topical today, with numerous television documentaries and news articles being broadcast in recent years focusing on the quality of both the resident and translocated fish populations from Loch Leven.

In the mid 1830s, the loch became an important water supply for downstream industry. As such, it was subject to a range of hydrological modifications. These included straightening the outflow to lower the level of the loch and the installing sluice gates to control the rate of discharge (Munro, 1994). The stated aims of these engineering works, which cost the equivalent of about $US 60,000 at the time (approximately $US 4.5 M by present day values), were:

1. to increase water availability by reducing the surface area and, consequently, evaporative losses from the surface of the loch
2. to create additional farmland around the edge of the loch
3. to control the outflow to meet the water requirements of downstream industry

The physical impact of this engineering work on the loch is well documented (Munro, 1994). The depth was reduced by 1.4 m, the surface area was reduced by 4.5 km$^2$, and the shoreline was moved by up to 500 m in places. In addition, the four islands within the loch were enlarged and three new islands were created. The final dimensions of the loch and its catchment are summarised in Table 1.

The changes in the hydromorphology of the system also affected the ecology of the loch. The records note that the installation of the sluice gates prevented salmon, sea trout and charr from entering the loch via the outflow, and that the contraction of the shoreline left established beds of aquatic plants above the waterline, leaving the margin of the loch unsuitable for supplying food for the fish (Fleming, 1836). However, the reduction in loch volume would also have increased the annual flushing rate by about 25%, which may have helped reduce the likelihood of algal blooms developing in later years when the loch started to become more eutrophic. At the end of the project the benefits of these works were reviewed and it was concluded that the predicted advantages had been exaggerated. Overall, the results were found to be disappointing; the land reclaimed for farming was poor, the value of this world famous fishery had been reduced by 1/3rd (Fleming, 1836) and the final cost was nine times that originally estimated (Munro, 1994).

**Table 1** Morphometric data for Loch Leven after the engineering works were completed in the mid 1830s (after Smith, 1974)

| | |
|---|---|
| Mean depth | 3.9 m |
| Maximum depth | 25.5 m |
| Surface area | 13.3 km$^2$ |
| Volume | 52.4 m$^3$ |
| Length | 5.9 km |
| Breadth | 2.3 km |
| Length of shoreline | 18.5 km |
| Shoreline development | 1.43 |

## Early research at the site

Loch Leven has been an important site for scientific research for almost 200 years, with papers on algae (Wesenburg-Lund, 1905; Bachmann, 1906; Rosenberg, 1938) macrophytes (Hooker, 1821; Scott, 1891; West, 1910), invertebrates (Scott, 1891, 1899, 1906), fish (Fleming, 1836; Burns-Begg, 1874; Johnstone, 1934; Rosenberg, 1938) and wildfowl (Sibbald, 1710; Smith, 1973; Buchanan, 1839; Millais, 1901, 1902, 1913) appearing in a range of scientific journals during the 1800s and early 1900s. The loch was also part of the bathymetric survey of Scotland carried out by Murray & Pullar (1910), which created a very detailed bathymetric map of the loch (Fig. 2) from 538 individual depth soundings that were taken using a bicycle wheel, some galvanized wire and a weight (Fig. 3).

## The International Biological Programme (IBP) and its importance

In 1966, Loch Leven was selected to be part of a large research programme known as the International Biological Programme (IBP). The IBP was designed to advance understanding of the effects of human population increase on the productivity, resilience and management of terrestrial, freshwater and marine ecosystems. This was to be achieved using international collaborative research. The key aims of this Programme were as follows:

- to apply "big science" to ecosystem ecology by studying ecological processes in lakes from the tropics to the arctic,
- to take a whole ecosystem approach to understanding lake ecology,

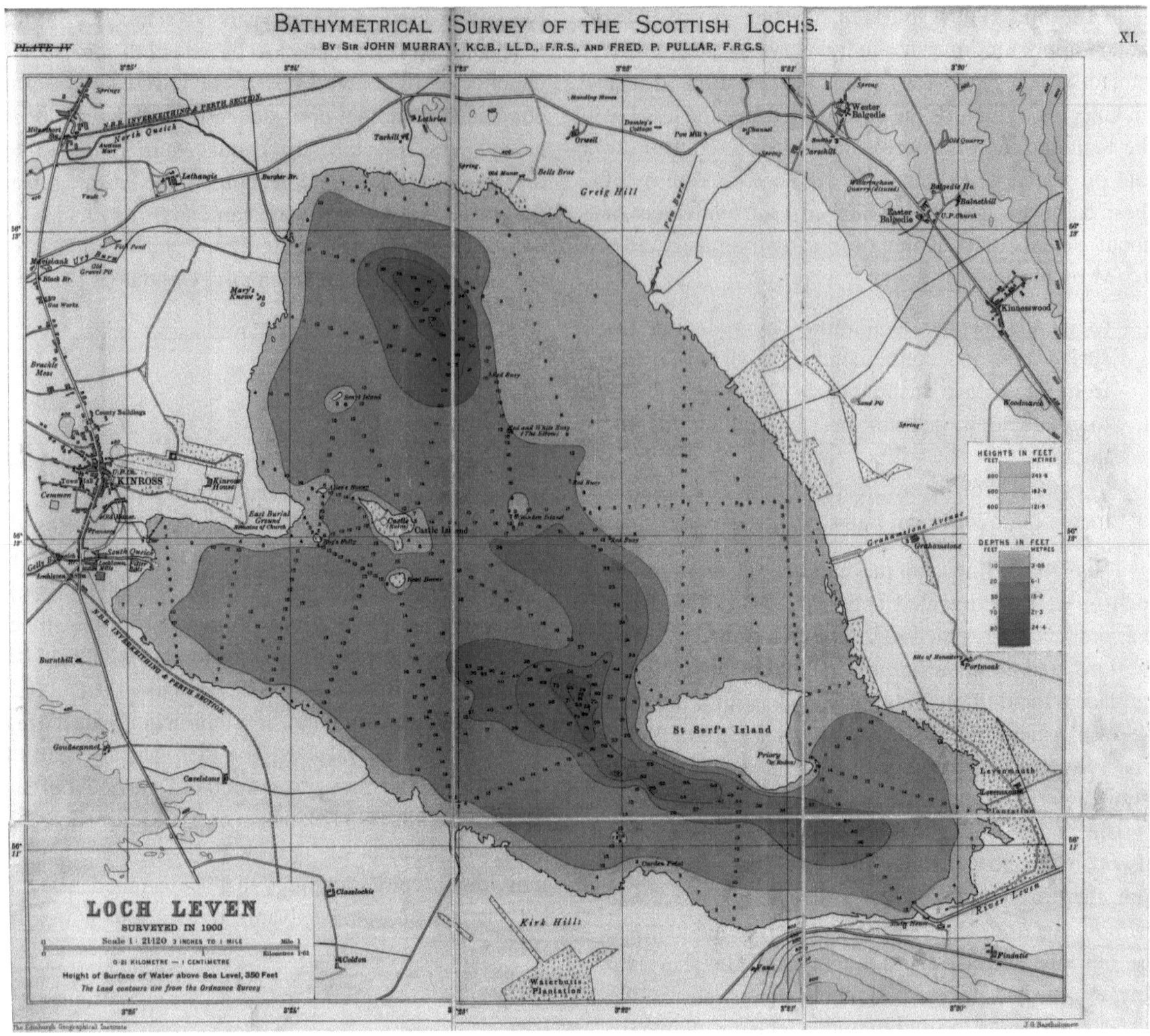


**Fig. 2** Bathymetric map of Loch Leven, 1900, showing depth sounding points on transects (after Murray & Pullar, 1910)

- to understand how biological systems respond to environmental change,
- to conserve natural resources for human benefit,
- to encourage man to consider the interests of ecology when making management decisions.

Loch Leven was selected as one of the main UK IBP sites and contributed to an impressive 43 lakes and 12 reservoirs from across the world that were included in the 'Productivity of Freshwater Communities Programme'. The work at Loch Leven was initially chaired by Sir Maurice Young (March 1966–1967) and later by Dr. J. W. G. Lund.

The IBP-funded work at Loch Leven continued until 1973 and involved large numbers of researchers. In one year alone, this amounted to 41 members of staff and 30 person years of effort. One of those researchers is shown in Fig. 4, collecting benthic invertebrate samples with a Jenkin surface mud sampler (Mortimer, 1942). The project culminated in a large symposium sponsored by the Royal Society of Edinburgh (Stirling University, 11–13 June 1973) and a number of published articles that documented the key findings of the research programme (Royal Society of Edinburgh, 1974). These articles pushed the boundaries of shallow lake research and included

**Fig. 3** **a** F. P. Pullar, **b** Sir John Murray on their last expedition to Loch Leven, 1 September 1900, and **c** their depth sounding equipment (after Murray & Pullar, 1910)

the results of investigations into the physical (e.g. bathymetry, hydrology and meteorology), chemical (including catchment, water-column and sediment characterisation) and biological (including birds, fish, plankton and the benthos) structure and function of the loch. Most striking were the linkages made between each of these areas in the summary paper by Morgan & McLusky (1974), in which a whole system energy transfer budget was presented. This ecosystem-based approach was both comprehensive and novel at the time of publication. The symposium volume concludes with some recommendations for future work, including:

- identifying the factors that limit phytoplankton production
- investigating ways of reducing phosphate input to the lake from industrial sources and sewage works
- reducing nitrate inputs from agricultural sources
- reducing internal nutrient recycling from the sediments
- re-establishing submerged macrophytes and emergent vegetation
- monitoring to detect significant changes that may require management intervention

All of these areas continue to drive research on Loch Leven, and many other freshwater lakes, to this day. More recently, however, the effects of climate change have also become the focus of attention. One of the most noticeable effects of this on the loch is higher water temperatures in winter. As a result,

**Fig. 4** Sampling for macroinvertebrates with a Jenkin surface mud sampler during the IBP

annual curling competitions are no longer possible because the loch has rarely frozen well enough to support such activities in recent years.

## Research at Loch Leven after the IBP

When the IBP drew to a close, the UK Natural Environment Research Council (NERC) continued to fund a research and monitoring programme at this site, albeit in a much reduced form. This has been maintained to the present day, resulting in a unique collection of long-term data that currently spans more than 40 years. The data are more or less continuous, although funding cuts in the mid 1980s have created some gaps in the data series. Complementary information collected by other organisations over this period and beyond (including daily outflow measurements since 1850, detailed fish catch records since 1900 and a range of macrophyte records dating back to 1821) adds considerable scientific value to this core dataset.

Collaboration has been pivotal to the success and development of the long-term monitoring programme. As a direct result, Loch Leven continues to be used as a platform for a wide range of academic and policy oriented collaborative research projects. Further, the site has been continually used as a training ground for students with research activities ranging from "pond-dipping" projects with local primary schools to the hosting of international postdoctoral research staff. The efforts taken by research staff to disseminate scientific findings to the public was recently acknowledged by the Scottish Parliament, when the "Growing Up With Loch Leven" initiative (led by Scottish Natural Heritage) was described as an example of "the benefits that Real World Learning could bring to the Scottish curriculum" (Dr. Sylvia Jackson, 14th February 2007; Scottish Parliament Business Bulletin No. 28/2007).

However, it is the almost unique co-operation between researchers, policy makers and other stakeholders at this site that maintains the international importance of Loch Leven. After the success of the IBP, a few high profile 'incidents' kept Loch Leven high on the research agenda and resulted in renewed interest in collaborative research at this site. Most prominent of these was the 'fall-out' from press coverage of a serious cyanobacterial bloom in the early 1990s, which included a debate in the UK parliament on water quality issues at Loch Leven and their wider implications. This, ultimately, led to an increase in research effort to identify the causes, effects and management of cultural eutrophication at the site (LLCMP, 1999), and Loch Leven now boasts one of the most comprehensively monitored examples of lake restoration (costing > $US6.3 M to date) in the world. This is still used as a 'best practice' case study in the development of other lake management plans. Other contentious issues have resulted in similar press coverage and governmental intervention, most notably the recent debate surrounding the apparent ecological conflicts between cormorants and fish, and the consequent socio-economic conflicts between fisheries managers and conservation authorities. Addressing these issues has improved scientific understanding of the ecosystem services provided by the loch and provided the evidence base that underpins the development and implementation of good management practices by local stakeholders.

In celebration of the recent 40th year of the long term monitoring programme, and to highlight the wide range of scientific investigations still being conducted at Loch Leven, the NERC Centre for Ecology & Hydrology (CEH) organised a symposium entitled "Loch Leven: 40 years of scientific research", which was held in Kinross, Scotland, UK, on 11 December 2008. The main theme of the symposium was the role of long term monitoring in developing our understanding of the links between pollution, climate change and ecological responses in shallow lakes. The results presented at the meeting are summarised in the series of articles that follow.

**Acknowledgements** We thank the Kinross Estates, Loch Leven Fisheries, Loch Leven Trustees, Scottish Natural Heritage, Scottish Environment Protection Agency and Royal Society for the Protection of Birds for their support for our scientific research at Loch Leven. We are also grateful to Alex Kirika and the late Tony Bailey-Watts who maintained the long-term monitoring programme at Loch Leven for more than 30 years. This research was funded by Scottish Natural Heritage, the Scottish Environment Protection Agency and the Natural Environment Research Council, UK. Loch Leven is part of the UK Environmental Change Network (http://www.ecn.ac.uk/).

## References


Bachmann, H., 1906. Le plankton Écossais. Archives des Sciences Physiques et Naturelles 20: 359.

Buchanan, G., 1839. Parish of Kinross. In The statistical account of Scotland, 1834–1845, volume 9: 2–29.

Burns-Begg, R., 1874. The Loch Leven Angler, Edinburgh and Glasgow. George Barnett, Kinross: 144 pp.

Fleming, J., 1836. Interim report by the Revd Dr Fleming as valuator of the damage done to the fishings by the drainage of Loch Leven, Kinross, MS. Sheriff Court, Kinross.

Hooker, W. J., 1821. Flora Scotica: or a description of Scottish Plants, London.

Johnstone, J. E., 1934. The Loch Leven Angler. Brown, Kinross: 148 pp.

LLCMP, 1999. Loch Leven Catchment Management Plan: 93 pp.

Millais, J. G., 1901. The Wildfowler in Scotland. Longmanns, London.

Millais, J. G., 1902. The Natural History of British Surface-feeding Ducks. Longmanns, London.

Millais, J. G., 1913. The British Diving Ducks. Longmanns, London.

Morgan, N. C., 1974. Historical background to International Biological Program Project at Loch Leven, Kinross. Proceedings of the Royal Society of Edinburgh Section (B) 74: 45–55.

Morgan, N. C. & D. S. McLusky, 1974. Summary of the Loch Leven IBP results in relation to lake management and future research. Proceedings of the Royal Society of Edinburgh Section (B) 74: 407–416.

Mortimer, C. H., 1942. The exchange of dissolved substances between mud and water in lakes. Journal of Ecology 30: 147–201.

Munro, D., 1994. Loch Leven and the River Leven: a landscape transformed. River Leven Trust, Markinch: 196 pp.

Murray, J. & L. Pullar, 1910. Bathymetrical Survey of the Freshwater Lochs of Scotland. Challenger Office, Edinburgh.

Rosenberg, M., 1938. Algae and trout. A biological aspect of the poor trout season in 1937. Salmon and Trout Magazine 89: 313–322.

Royal Society of Edinburgh, 1974. The Loch Leven I.B.P. Project. Proceedings of the Royal Society of Edinburgh Section (B) 74: 43–421.

Scott, T., 1891. The invertebrate fauna of the inland waters of Scotland, Part II. In 9th Report of the Fishery Board of Scotland Part III: 269–296.

Scott, T., 1899. The invertebrate fauna of the inland waters of Scotland—report on special investigation. In 17th Report of the Fishery Board of Scotland Part III: 161–165.

Scott, T., 1906. A catalogue of land, fresh-water, and marine Crustacea found in the basin of the River Forth and its estuary. Proceedings of the Royal Physical Society of Edinburgh 16(97–190): 267–386.

Sibbald, R., 1710. The History, Ancient and Modern, of the Sherrifdoms of Fife and Kinross, Edinburgh.

Smith, I. R., 1973. The assessment of winds at Loch Leven, Kinross. Weather 28: 202–210.

Smith, I. R., 1974. The structure and physical environment of Loch Leven, Scotland. Proceedings of the Royal Society of Edinburgh Section, B 74: 81–100.

Wesenburg-Lund, C., 1905. A comparative study of the lakes of Scotland and Denmark. Proceedings of the Royal Society of Edinburgh, B 25: 401–448.

West, G. S., 1910. A further contribution to a comparative study of the dominant Phanerogamic and Higher Cryptogamic flora of aquatic habit in Scottish lakes. Proceedings of the Royal Society of Edinburgh, B 30: 170–173. 62 Plates.

Yarrell, W., 1839. Supplement to the second volume of the History of British Fishes. Samuel Bentley, London: 9–12.

Hydrobiologia (2012) 681:11–21
DOI 10.1007/s10750-011-0922-y



# Historical changes (1905–2005) in external phosphorus loads to Loch Leven, Scotland, UK

L. May · L. H. Defew · H. Bennion · A. Kirika

Published online: 9 November 2011


**Abstract** This article reviews historical changes in the total phosphorus (TP) inputs to Loch Leven, Scotland, UK. Data derived from palaeolimnological records suggest that inputs in the early 1900s were about 6 t TP year$^{-1}$ (0.45 g TP m$^{-2}$ year$^{-1}$). By 1985, this had risen to about 20 t TP year$^{-1}$ (1.5 g TP m$^{-2}$ year$^{-1}$) due to increases in runoff from agricultural land and discharges from point sources. By the late 1970s, increased TP inputs were causing serious degradation of lake water quality. Most noticeably, there had been an increase in cyanobacterial blooms. A catchment management plan was implemented in the early 1990s. This resulted in a 60% reduction in the annual TP input between 1985 (20 t TP year$^{-1}$/1.5 g TP m$^{-2}$ year$^{-1}$) and 1995 (8 t TP year$^{-1}$/0.6 g TP m$^{-2}$ year$^{-1}$). The main reduction was associated with better control of point source discharges, but attempts were also made to reduce inputs from diffuse sources. The reduction in external TP loading to the lake led to a marked decline in TP retention by the lake each year.

**Keywords** Lake · Eutrophication · Restoration

Guest editors: L. May & B. M. Spears / Loch Leven: 40 years of scientific research

L. May (✉) · L. H. Defew · A. Kirika
Centre for Ecology & Hydrology, Bush Estate, Penicuik, Midlothian EH26 0QB, UK
e-mail: lmay@ceh.ac.uk

H. Bennion
Environmental Change Research Centre, Department of Geography, University College London, Gower Street, London WC1E 6BT, UK

## Introduction

External nutrient loading has long been recognised as one the most important factors controlling the productivity or trophic state of a lake (Vollenweider, 1968; Dillon & Rigler, 1974; Jeppesen et al., 2005). Under natural conditions, nutrient inputs to lakes are generally low and cause few water quality problems. However, anthropogenic activity within the catchment can increase nutrient inputs to waterbodies to a level that degrades water quality (Gulati & Van Donk, 2002) and promotes troublesome, and sometimes toxic, cyanobacterial blooms (Codd, 1995; Schindler, 2006; Leigh et al., 2010). This process is known as eutrophication. Such activities include fertiliser applications to farmland, animal husbandry, population increases and industrial development (Harper, 1992; Jeppesen et al., 2007).

Eutrophication is the most widespread water quality problem in many parts of the world (Carpenter et al., 1998) and dealing with it can be costly. For example, Pretty et al. (2003) estimated that responding to eutrophication problems in England and Wales, alone, cost c. USD 89M year$^{-1}$, with associated social and ecological damage estimated to be a further USD

121M–232M year$^{-1}$. In addition to these short-term costs, there are also enormous additional costs associated with the restoration of lakes that have undergone cultural eutrophication to achieve the 'good' or 'high' ecological status required by the EU Water Framework Directive (European Parliament, 2000). For this reason, it is important that the most cost effective solutions to lake eutrophication problems are used and these need to be derived from the results of well-documented, previous, restoration projects (Sondergaard & Jepessen, 2007).

Loch Leven (Scotland, UK) is a good example of a large, shallow lake with a long, and well documented, history of water quality problems caused by anthropogenic eutrophication. The most noticeable symptom of these problems was an increase in the frequency of cyanobacterial blooms from 1963 onwards (Holden & Caines, 1974; Morgan, 1974). By the mid-1980s, these blooms had become so serious that they threatened the high conservation, recreational and economic value of the lake (May & Spears, 2011a). As early as the 1970s, it was broadly recognised that this degradation in water quality was closely related to increases in nutrient-laden agricultural runoff and elevated levels of discharges from waste water treatment works (WWTWs) and industrial sources (Holden & Caines, 1974).

By the 1980s, it was generally believed that phosphorus (P) was the main nutrient limiting phytoplankton production and biomass accumulation in most shallow lakes (e.g. Schindler, 1977). Long-term water quality records from Loch Leven were examined and these showed a close, positive relationship between algal abundance (expressed as chlorophyll *a* concentration) and total P (TP) availability (Fig. 1). It was, therefore, concluded that reducing TP inputs to the lake would improve water quality and this became the focus of a range of catchment management activities undertaken in the late 1980s and early 1990s (LLCMP, 1999).

Increased TP input associated with anthropogenic development within the catchment seemed to be the main cause of the eutrophication problems at Loch Leven. Initial estimates suggested that, between 1967 and 1976, the external TP input to the lake ranged between 7 and 17 t year$^{-1}$(0.53–1.28 g m$^{-2}$ year$^{-1}$), with up to 70% of that input coming from a single industrial source (Holden & Caines, 1974; Holden et al., 1975; Caines & Harriman, 1976). By 1985, the

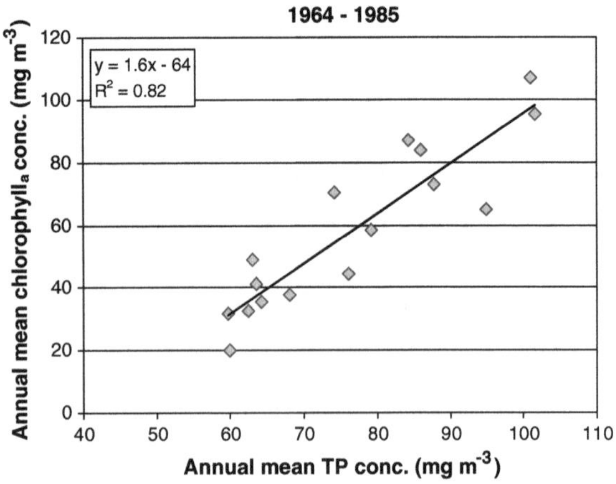


**Fig. 1** The relationship between annual mean in-lake TP concentration and annual mean algal abundance (expressed as chlorophyll *a* concentration) in Loch Leven, 1964–1985 ($P < 0.001$)

overall TP input had risen to about 20 t year$^{-1}$ (1.5 g m$^{-2}$ year$^{-1}$), 57% of which could be attributed to point source discharges from WWTWs and industry (Bailey-Watts & Kirika, 1987; Bailey-Watts et al., 1987).

Between 1985 and 1995, controls imposed on industrial discharges from a local woollen mill (D'Arcy, 1991) and improvements to local sewage treatment facilities (Bailey-Watts & Kirika, 1996, 1999) reduced the TP input to the lake from point sources by about 70% (Bailey-Watts & Kirika, 1996, 1999). However, a 53% fall in TP transported by runoff between 1985 and 1995 also reduced the overall TP load (Bailey-Watts & Kirika, 1999). At the end of this period, the external TP load to the lake was estimated to be about 8 t year$^{-1}$ (0.6 g m$^{-2}$ year$^{-1}$). It should be noted, however, that the reduction in runoff may have been caused by inter-annual variation in rainfall, which was about 30% lower in 1995 than in 1985, rather than changes in catchment management activities (Bailey-Watts & Kirika, 1996, 1999).

Management activities aimed at reducing the TP input to the lake continued beyond 1995 as part of a long-term strategy for its restoration (LLCMP, 1999). Although at a much lower level than before, and focusing on reducing TP losses from diffuse sources rather than point sources, these activities aimed to provide a sustainable future for the lake. Most activities were focused on one particular feeder stream, the Pow Burn and its catchment (Fig. 2).

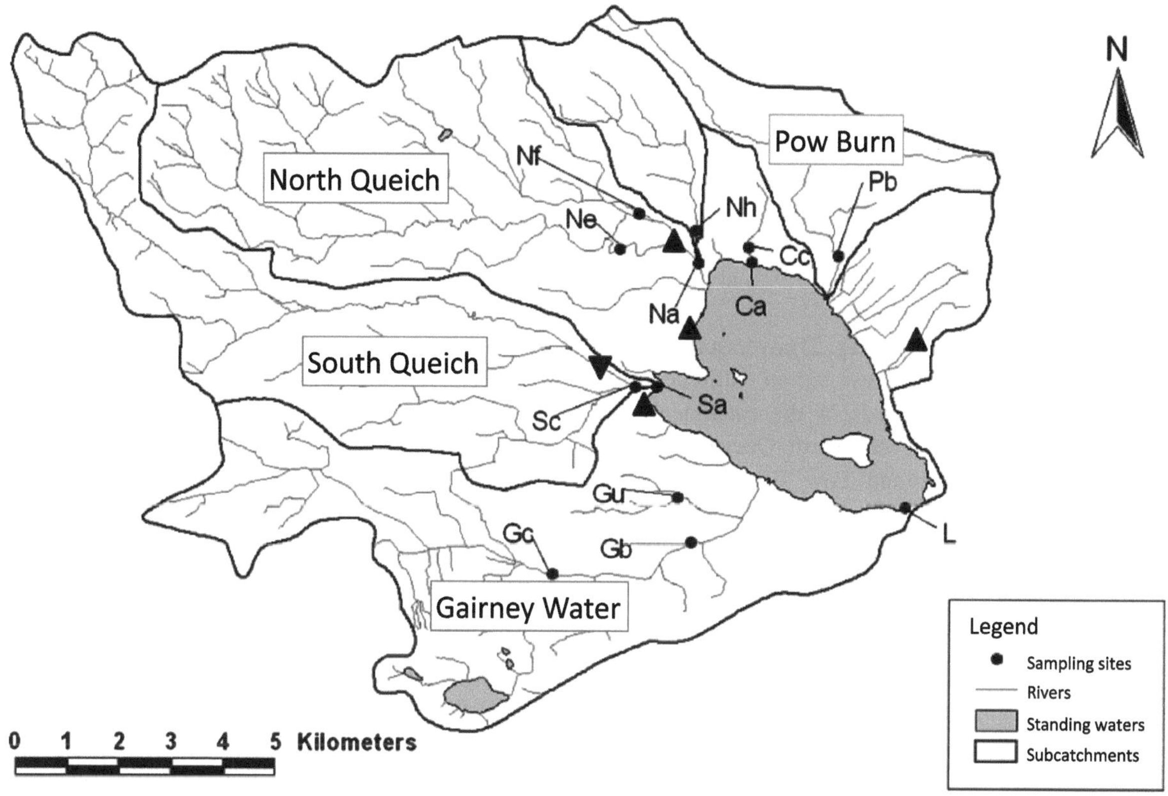


**Fig. 2** Map of the Loch Leven catchment showing sampling sites on the main inflows [Pow Burn (P), North Queich (N), South Queich (S), Gairney Water (G)] and at the outflow (L) in 2005 and corresponding catchments for each stream; the position of point source discharges in 1985 are indicated (waste water treatment works: *triangles*; industry: *inverted triangle*)

Although only accounting for about 10% of the lake's catchment, this stream had been shown to be the source of about 30% of the total river-borne P entering the lake during 1985 (Bailey-Watts & Kirika, 1987). Fields within this sub-catchment had also been identified as being at high risk of soil erosion (Frost, 1994).

In-field land management techniques were introduced to reduce the risk of soil erosion at source within the Pow Burn catchment and a series of grass or grass/tree buffer strips up to 20 m wide were installed along the banks of this inflow and its feeder streams to intercept eroded material and prevent it from entering the watercourses (Castle et al., 1999). Although there is little evidence that these buffer strips significantly reduced P runoff from this area under 'normal' conditions, Vinten et al. (2004) concluded that they may have been at least partially effective during high runoff events. In addition to these specific management activities, local farmers were encouraged to control livestock grazing, fertiliser usage and slurry/manure spreading across the whole of the catchment (Castle et al., 1999).

The main aim of this article is to review recent historical changes in the external TP input to Loch Leven, using data compiled from published records for the period 1965 to 1995 and new data from 2005, and compare them with 'baseline' data derived from palaeolimnological records. Changes in the annual TP balance of the lake between 1975 and 2005 are explored and the effectiveness of catchment management measures at reducing the TP input to the lake is considered. This article also provides background information and context for the series of papers that follow, which describe the ecological response of the lake to these historical changes in nutrient conditions (Carss et al., 2011; Carvalho et al., 2011; Dudley et al., 2011; Elliott & Defew, 2011; Gunn et al., 2011; May & Spears, 2011b; Spears et al., 2011; Winfield et al., 2011).

## Site description

Loch Leven is a shallow lake situated in the lowlands of Scotland, UK. The structure and physical environment of the lake are described in detail by Smith (1974). In summary, the lake lies at a latitude of 56°10′N, a longitude of 3°30′W and an altitude of 107 m.a.o.d. It has a surface area of 13.3 $km^2$, and mean and maximum depths of 3.9 and 25.5 m, respectively (Kirby, 1971).

The catchment of the lake (Fig. 2) covers an area of about 145 $km^2$, which rises to an altitude of 482 m.a.o.d. About two-thirds of the catchment is drained by four main rivers (North Queich, South Queich, Gairney Water and Pow Burn), and the remainder is drained by several minor inflows and some small areas of land along the shoreline that drain directly into the lake (Fig. 2). The catchment is mainly rural and intensively farmed, with about 80% being used for agricultural production (LLCMP, 1999). Currently, the land use in this area is dominated by mixed farmland with the best land being used for high-value potato and vegetable crops, cereals and oilseed rape (Castle et al., 1999). There is some livestock rearing, but this is mainly restricted to the more upland areas, which tend to be further away from the lake (LLCMP, 1999). A further 11% of the catchment is woodland and about 2% is used for habitation. High levels of P-laden runoff enter the lake from areas of intensive agriculture within the catchment.

Population density in this area is low, with only about 11,000 people living within the catchment (Frost, 1996). About 60% of these live in the towns of Kinross, Milnathort and Kinnesswood (Perth & Kinross Council, 2004), which are served by mains sewerage networks that are connected to waste water treatment works (WWTWs). However, it is estimated that about 650 households in the more remote parts of the catchment rely upon private sewage treatment systems, such as septic tank systems, for managing their domestic waste (Dudley & May, 2007). In the past, high levels of P have entered the lake in the form of WWTW effluent (Bailey-Watts & Kirika, 1999) and septic tanks discharges (Frost, 1996).

Although there is little industry within the catchment, there have been one or more woollen mills on the banks of one of the South Queich since 1840 (Munro, 1994). These mills have discharged large quantities of P-laden effluent into the lake in the past, but the introduction and enforcement of strict discharge limits in the late 1980s reduced TP discharges from the one remaining mill from about 6 t TP $year^{-1}$ (0.45 g TP $m^{-2}$ $year^{-1}$) in 1985 to about 0.4 t TP $year^{-1}$ (0.03 g TP $m^{-2}$ $year^{-1}$) by 1988 (D'Arcy, 1991; D'Arcy et al., 2006). More recently, processing methods at the mill have changed and there is no longer any discharge of TP to the lake from this source (D'Arcy, 1991).

## Methods

### Historical TP loads, 1905–1965

Inputs of TP to the lake for the period 1905–1965 were derived from palaeolimnological records, because no measured values were available for this period. This was achieved by, first, inferring in-lake TP concentrations from the diatom assemblages in a 90 cm sediment core (LEVE11) that was taken with a mini-Mackereth corer (Mackereth, 1969) on 5-5-1999. The core was collected at a water depth of 4 m near to the Reed Bower site and extruded at 1 cm intervals.

The core was radiometrically dated using standard gamma assay procedures (Appleby et al., 1986, 1992) and the chronology was calculated based on $^{210}Pb$ using the constant rate of supply model (Appleby & Oldfield, 1978). Sediment accumulation rates of 0.12–0.15 cm $year^{-1}$ were estimated for this core for the period ~1930–1970, and a rate of 0.12 cm $year^{-1}$ was assumed for the period prior to 1930. The errors associated with the dating were relatively small, with errors of ±5–10 years for the period 1950–1970, and ±10–14 years for the period 1930–1950. Sixteen sub-samples were prepared and analysed for diatoms using standard procedures (Battarbee et al., 2001).

The reconstruction of diatom-inferred TP (DI-TP) was produced using a northwest European training set of 152 relatively small, shallow lakes (<10 m maximum depth) with a median value for the data of 104 μg TP $l^{-1}$ and a root mean squared error of prediction (RMSEP) of 0.22 $\log_{10}$ μg TP $l^{-1}$ for the weighted averaging model (Bennion et al., 1996, 2004). The reconstructions were implemented in C2 (Juggins, 2003), following taxonomic harmonization between the training set and the fossil data.

The DI-TP values were used to back-calculate the corresponding TP load to the lake using the equation of Dillon & Rigler (1974) and annual flushing rates calculated from daily outflow records (Sargent & Ledger, 1992; Loch Leven Trustees, pers. comm.). The resultant annual TP loading values were averaged over five 10-year periods, i.e. 1900–1909, 1910–1919, 1930–1939, 1950–1959 and 1960–1969, to take account of uncertainties in the core chronology. Each derived value was assigned to the 5th year of each decade, i.e. 1905, 1915, 1935, 1955 and 1965.

TP load for 2005

The TP load to Loch Leven in 2005 was estimated from measured in-stream nutrient concentrations and measured or derived flows determined for each of the 12 major inflows and the outflow at 8-day intervals between 18 January 2005 and 17 January 2006 (Fig. 2). Water samples were collected with 250 ml plastic sample bottles that had been pre-washed in Decon 90®, rinsed with distilled water and then rinsed again with stream water prior to sample collection.

Duplicate stream water samples were collected at each site by submerging the sample bottles to about 20 cm below the water surface. Samples were analysed for TP content using the methods of Murphy & Riley (1962), as modified by Eisenreich et al. (1975), on samples that had been subjected to an acid hydrolysis reaction prior to colorimetric analysis. This was achieved by mixing the sample with a solution of sulphuric acid ($H_2SO_4$) and potassium persulfate ($K_2S_2O_8$) and then autoclaving at 121°C for 30 min. The intensity of colour formed was quantified by measuring absorbance with a Philips PU8620 Series UV/VIS/NIR single-beam spectrophotometer, using a 4 cm flow cell and a wavelength of 882 nm. The concentration of TP within each sample was determined from this value using a linear calibration equation generated from the measured absorbance values of known TP standards. All analyses were performed in triplicate to ensure accuracy and quality of the results. The TP detection limit was estimated to be 2 $\mu g\ l^{-1}$.

Flow data from continuous recording devices located on the Pow Burn (Pb), South Queich (Sc) and North Queich (Ne) were provided by the Scottish Environment Protection Agency (SEPA). This information was supplemented by water height measurements recorded from gauging boards on the smaller streams, i.e. Camel Burn (Cc), Gairney Water (Gb) and the Hatton Burn (Nh), on each sampling date. Stream flows were estimated from these water height measurements using ratings curves that had been constructed from a range of measured in-stream flows across known gauge heights (Defew, 2008).

Two small streams flowing into the lake, i.e. Ury Burn and Kinnesswood Burn, were not included in the field survey. Phosphorus concentrations at these sites were assumed to be similar to those of the nearest sampled sub-catchment, i.e. the North Queich and the Pow Burn, respectively. Stream flow at these un-gauged sites was assumed to be directly proportional to the measured stream flows at the nearby North Queich and the Pow Burn in terms of relative catchment area, because Bailey-Watts and Kirika (1987) had previously found strong correlations ($r = 0.97$, TP $< 0.05$) between mean discharge and drainage area at these sites. This proportional method was also used to account for additional stream flow downstream of sampling sites that were not close to the confluence of the stream with the lake, and for water draining directly into the lake from areas outside the sampled sub-catchments.

The annual TP load to the lake from each inflow or area of direct runoff was calculated using the method of Stevens & Smith (1978), as follows:

$$\text{Total Load} = K\left(\sum_{i=1}^{n}(C_c Q_c)\right)$$

where $K$ is a factor to convert from time period of record to annual value, $n$ is number of samples, $C_c$ is estimate of TP concentration from continuous stream flow value and $Q_c$ is the measured or estimated flow ($m^3\ s^{-1}$).

The estimated TP loads were corrected statistically following the method of Ferguson (1986) and the final values were summed to give an overall annual TP load to the lake from its catchment. Further details of these methods are given by Defew (2008).

As P loads to the lake also include inputs from non-catchment sources such as excreta from over-wintering geese and rain falling directly onto the surface of the lake, these were also estimated. Phosphorus inputs from geese were calculated from data provided by Scottish Natural Heritage which suggested that, on average, 12,656 pink-footed geese per day roosted on the lake from January to March and between September and December, 2005. Assuming that these geese were feeding, primarily, on grass, the input per goose was estimated to be c. 172 mg TP $d^{-1}$ (Hancock, 1982). Direct inputs from rainfall were calculated from the total amount of rain falling onto the lake surface each year (data provided by Kinross Estates) and the estimated TP content of rain in this area, i.e. c. 25 mg TP $m^{-3}$ $year^{-1}$(Bailey-Watts et al., 1987).

Phosphorus retention rates

Annual TP retention rates for Loch Leven were calculated at 10-yearly intervals for the years for which detailed TP loading data were available, i.e. 1975, 1985, 1995 and 2005, using a mass balance approach (Table 1). This involved calculating the annual discharge of TP from the lake for each year from the TP concentrations measured at or near the outflow at roughly weekly intervals (Caines & Harriman, 1976; Bailey-Watts & Kirika, 1987, 1996, 1999; Bailey-Watts et al., 1987; this study) and the corresponding rates of hydraulic discharge (data provided by the Loch Leven Trustees). In outline, discharges of TP from the outflow on each sampling occasion (kg $day^{-1}$) were estimated as the product of the corresponding concentration and flow values outlined above. The average of these daily values was then multiplied by 365 to estimate the discharge of TP from the lake in each year. The TP retention coefficient for the lake was then calculated by subtracting the annual TP discharge value from the corresponding annual TP loading value for each year, dividing the result by the annual TP loading, and then expressing the result as a percentage.

**Table 1** Point sources of total phosphorus (TP), and the size and timing of significant reductions in their respective output, within the catchment of Loch Leven between 1985 and 2005 (after LLCMP, 1999)

| Point source | Anticipated reduction in TP output (t $year^{-1}$/g $m^{-2}$) | Date achieved by |
|---|---|---|
| Woollen mill | 6.3/0.47 | 1987 |
| Kinross WWTW | 1.7/0.13 | 1993 |
| Kinross WWTW | 0.4/0.03 | 1997 |
| Milnathort WWTW | 0.6/0.05 | 1995 |
| Kinnesswood WWTW | 0.6/0.05 | 1997 |
| Total reduction | 9.6/0.72 | 1997 |

*WWTW* waste water treatment works

## Results

The diatom-inferred TP concentration (DI-TP) in Loch Leven was relatively stable for the period 1905–1965 at ~65–75 μg TP $l^{-1}$. The DI-TP values for the uppermost part of the core were in very good agreement with the measured 3-year mean TP concentration of the lake for 1997–1999 suggesting that the reconstructions are likely to be reliable. Furthermore, comparison of DI-TP values from a more recent core with measured annual mean TP concentrations for the period 1969–2005 shows that the DI-TP reconstruction agrees well with the time series of measured values in terms of both the overall trend and the actual TP values (Bennion et al., unpublished).

The annual TP inputs to Loch Leven between 1905 and 1965, as estimated from the palaeolimnological record, are shown in Fig. 3. Although the data suggest that the TP load was relatively stable over this period, i.e. about 6 t $year^{-1}$(0.45 g $m^{-2}$), it should be noted that this apparent stability may be an artefact generated by the decadal averaging method applied to the palaeolimnological data.

Measured data for the period between 1965 and 1995 showed that the annual TP load to the lake varied significantly over that period (Fig. 3), ranging from about 7.5 t $year^{-1}$ (0.56 g $m^{-2}$ $year^{-1}$) in the 1970s to about 20.5 t $year^{-1}$ (1.54 g $m^{-2}$ $year^{-1}$) by 1985, then falling to about 8 t $year^{-1}$ (0.6 g $m^{-2}$ $year^{-1}$) by 1995.

The most recent loading survey suggested that the TP load to the lake in 2005 was about 8.2 t $year^{-1}$

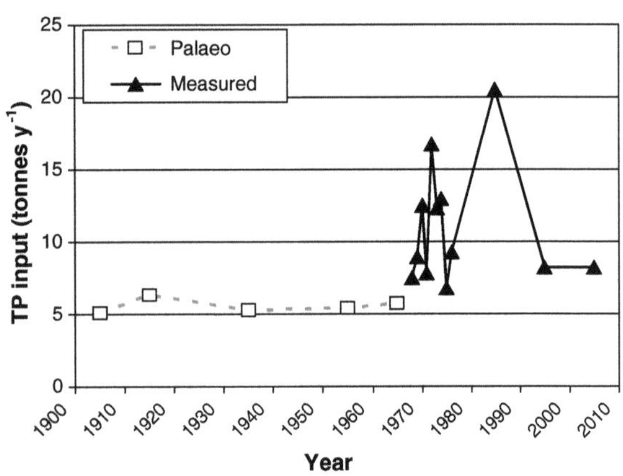


**Fig. 3** Historical changes in the external input of total phosphorus (TP) to Loch Leven between 1905 and 2005. *Open squares* are values calculated from palaeolimnological data (see text for details); *filled triangles* are measured values

(0.62 g m$^{-2}$ year$^{-1}$) (Fig. 3), with less than 6% of the external load being attributable to inputs from over-wintering geese and less than 5% being accounted for by rain falling directly onto the surface of the lake. This overall value was only about 2 t year$^{-1}$ (0.15 g m$^{-2}$ year$^{-1}$) greater than the average historical load that had been estimated for 1905–1965 and was, essentially, the same as that measured in 1995.

Changes in the calculated annual TP balance of the lake at 10-yearly intervals between 1975 and 2005 are shown in Fig. 4. Varying proportions of the TP entering the lake were retained in all of the years examined. In 1975, 44% of the external TP input (3 t year$^{-1}$/0.23 g m$^{-2}$ year$^{-1}$) was retained. By 1985, this had risen to 61% (12.5 t year$^{-1}$/ 0.94 g m$^{-2}$ year$^{-1}$) of the TP input. Once external inputs from the catchment had been reduced by about 60%, TP retention rates fell dramatically to 12% (1 t year$^{-1}$/0.08 g m$^{-2}$ year$^{-1}$) in 1995 and 15% (1.25 t year$^{-1}$/0.09 g m$^{-2}$ year$^{-1}$) in 2005. In contrast, however, the actual amount of TP discharged via the outflow was relatively constant each year, i.e. 1975: 4 t year$^{-1}$ (0.3 g m$^{-2}$ year$^{-1}$); 1985: 8 t year$^{-1}$ (0.6 g m$^{-2}$ year$^{-1}$); 1995: 7 t year$^{-1}$ (0.53 g m$^{-2}$ year$^{-1}$); 2005: 7 t year$^{-1}$ (0.53 g m$^{-2}$ year$^{-1}$) (Table 1).

## Discussion

The survey data from the studies reviewed above, and the additional data presented from this study, show that the TP load to Loch Leven increased from about 6 t year$^{-1}$(0.45 g m$^{-2}$ year$^{-1}$) to about 20 t year$^{-1}$(1.5 g m$^{-2}$ year$^{-1}$) over the main period of

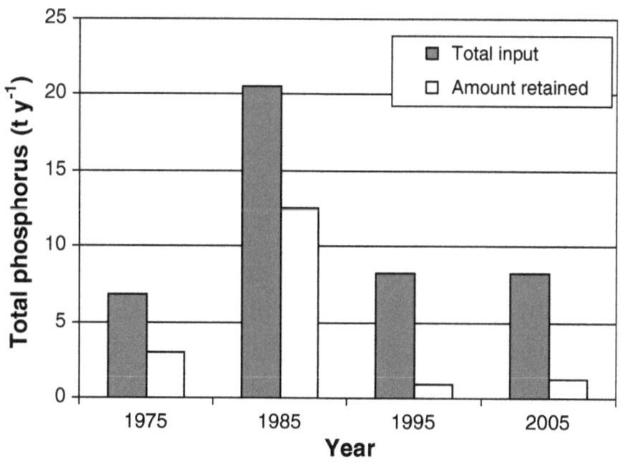


**Fig. 4** TP input and retention at Loch Leven estimated at decadal intervals from 1975 to 2005

enrichment. When catchment management measures were introduced in the late 1980s and early 1990s, this external load fell by about 12 t P year$^{-1}$(0.9 g P m$^{-2}$ year$^{-1}$), or 60%. This reduction comprised about 8.5 t P year$^{-1}$ (0.64 g P m$^{-2}$ y$^{-1}$) from point sources and about 3.5 t P year$^{-1}$ (0.26 g P m$^{-2}$ year$^{-1}$) from diffuse sources, although the latter is believed to be better explained by the lower rainfall in 1995 compared with 1985 (Bailey-Watts & Kirika, 1999) than by any improvements in catchment management. Since 1995, strict controls have continued to be enforced within this P sensitive catchment and, in 2005, P inputs to the lake were still low (i.e. 8 t year$^{-1}$/ 0.6 g m$^{-2}$ year$^{-1}$). This value is lower than the target value of 10 t year$^{-1}$(0.75 g m$^{-2}$ year$^{-1}$) set by the catchment management group (LLCMP, 1999) and seems to be approaching that estimated for the early 1900 s (i.e. 6 t year$^{-1}$/0.45 g m$^{-2}$ year$^{-1}$). The level of reduction of P input to Loch Leven is similar to that of other lakes that have undergone nutrient input reductions as part of lake restoration strategies. These include a 50% reduction in P load to Muggelsee, Germany (Kohler et al., 2005) and a 55% reduction in P load to Lake Apopka, USA (Coveney et al., 2005). An improvement in water quality followed in many of these lakes, albeit often following a significant time lag (i.e. 10–15 years: Jeppesen et al., 2005, 2007). However, this is not always the case. Gulati & van Donk (2002), in their review of the eutrophication of Dutch lakes and the restoration techniques applied, found that decreasing P inputs did not always result in an improvement in water quality.

When reducing the P input to a lake, it is important to identify the main sources of those inputs. At Loch Leven, when inputs were high, 56% of those inputs came from point source discharges with less than 44% coming from diffuse sources (Bailey-Watts & Kirika, 1987). Of the diffuse sources, P entering from rain falling directly onto the surface of the lake was generally low, i.e. about 0.3–0.4 t year$^{-1}$, which equated to about 2% of the external TP load in 1985 and 5% in 1995 and 2005. Inputs from geese over-wintering at the lake were also low, i.e. 0.5 t P year$^{-1}$, or about 3% of the external TP load in 1985 and about 6% of that load in 1995 and 2005. This is in contrast to the situation in many other shallow lakes where P inputs from wildfowl are relatively high (e.g. 70%: Manny et al., 1994; 88–92%: Ronicke et al., 2008; 73%: Chaichana et al., 2010). It is also likely that P inputs from geese to Loch Leven have changed very little in recent years because, although the proportion of greylag and pink footed has changed over time, the total number of geese has remained fairly stable (Carss et al., 2011).

There have been varying levels of TP retention within the lake over the period of available data, 1975–2005. However, on an annual timescale, P retention has remained positive, suggesting that TP continues to accumulate in the lake sediments. Such accumulations are important in terms of lake recovery because, when external TP loads are reduced, these internal sources tend to release P into the water column causing a considerable delay in recovery (Sas, 1989; Gulati & van Donk, 2002; Søndergaard et al., 2003; Nürnberg & LaZerte, 2004; Jeppesen et al., 2005; Søndergaard et al., 2007; Spears et al., 2011). That said, in the longer term, P release from the sediments also promotes recovery because it increases the rate at which TP from internal sources can be exported from the system, especially during the summer months (Sondergaard et al., 2001). In Loch Leven, however, hydrological discharge from the outflow is restricted during the summer months by the use of sluice gates that are employed to maintain a constant water supply for downstream users (May & Spears, 2011b). So, it is likely that P leaves this system more slowly than in naturally flushed lakes, thus slowing down the recovery process.

Data from Loch Leven for 1975 and 1985 show that, prior to the 60% reduction in external TP load, the TP retention rate was about 50%. After the reduction in external input, the TP retention rate fell to about 15%. Given that TP retention in lakes tends to be negatively correlated with hydraulic retention time (Brett & Benjamin, 2008), and that the latter 2 years were much drier than the earlier years, this suggests that that the lake is now retaining significantly less TP from external sources than it was in the 1970s and 1980s. However, total (rather than relative) export of TP from the lake has changed very little from year to year. This is probably because the rate of flow in the outflow, which is managed to ensure a constant flow of water to downstream industry over the summer period (May & Carvalho, 2010; May & Spears, 2011b), and in-lake annual average TP concentrations have remained fairly constant over this period (Carvalho et al., 2011). Although similar in terms of values, however, it should be noted that in-lake TP concentrations in the earlier years were driven by external TP loads, while those from 1995 onwards were increasingly driven by internal TP loads (Spears et al., 2011).

Although some relatively minor point source upgrades were implemented between 1995 and 2005, improvements in catchment management over that period were mainly focused on reducing TP inputs from agricultural (diffuse) sources through the better management of farm nutrient budgets and the installation of buffer strips along the banks of some of the inflows to reduce stream contamination by nutrient-laden runoff. Although a comparison of the TP inputs with the lake in 1995 and 2005 seems to suggest that these activities had little effect, the real impact of these measures cannot be determined by simply comparing these values directly. Source apportionment needs to be taken into account before conclusions are reached. This is not possible with the existing data, because the less detailed survey that was carried out in 2005 did not include sampling up- and down-stream of point source discharges. Nevertheless, it seems likely that the implementation of buffer strips, lower fertiliser application rates and other agricultural initiatives implemented across the catchment to reduce P losses from agriculture were probably more effective than the data suggest. This is because any resultant reduction in TP delivery to the lake from diffuse sources was probably masked by parallel increases in P-laden sewage effluent associated with recent population growth in this area. Government census statistics suggest that the resident population in the area around Loch Leven has grown rapidly in recent years, with an increase of about 22% occurring between 1981 and 1991 and 16% between 1991 and 2001 (Perth & Kinross Council, pers. comm.).

## Conclusion

This study shows that the implementation of a catchment management plan in the late 1980s and early 1990s reduced the external TP load to Loch Leven by about 60%. Most of this reduction was due to better control of point source discharges rather than reductions in nutrient-laden runoff from the surrounding farmland. Very little of the observed reduction could be attributed directly to the implementation of best management practices on agricultural land or the installation of buffer strips next to drainage channels.

However, this does not necessarily mean that these measures were ineffective. It is possible that any decreases in P-laden runoff from agricultural activities may have been masked by concurrent increases in population density and, consequently, higher TP discharges from WWTWs. This highlights the importance of obtaining good source apportionment data when assessing the effectiveness of catchment management activities in reducing nutrient loads to waterbodies.

When the input of TP to the lake from external sources was reduced by 60%, the overall TP retention rate fell from 50% to about 15%. This was because the rate of hydraulic discharge rate was unchanged and in-lake P concentrations remained high (Carvalho et al., 2011) as a result of increased P release from the sediments. This seemed to be 'compensating' for the reduction in TP load from the catchment (Spears et al., 2011) as has been observed in other lakes (Søndergaard et al., 2003; Nürnberg & LaZerte, 2004). This suggests that, while internal recycling appears to delay recovery in terms of in-lake water quality parameters, it also promotes recovery in terms of increasing the rate at which TP from internal sources is exported from the system. This should be taken into account when assessing the success of reductions in external load on the long-term recovery of eutrophic lakes.

**Acknowledgements** The authors are grateful to Willie Wilson of the Kinross Estates for providing daily rainfall data, the Loch Leven Trustees for daily outflow records, the Freshwater Fisheries Services at Pitlochry for phosphorus loading data from the 1970s and to the late Tony Bailey-Watts for phosphorus loading data from 1985 to 1995. The authors thank Scottish Natural Heritage (SNH) for information on over-wintering geese numbers, the Scottish Environment Protection Agency (SEPA) for flow records for 2005, and to Sandra Thompson of Perth & Kinross Council for population census data from the Kinross area for 1981, 1991 and 2001. The authors also thank Kate Heal for co-supervising the 2005 phosphorus loading study and we are indebted to the Montgomery family, the owners of the loch, for their continuing support for our research. The authors also thank two anonymous reviewers who helped to improve the manuscript. This research was mainly funded by the Natural Environment Research Council, UK, with some small contributions from SNH and SEPA. Loch Leven is part of the UK Environmental Change Network (http://www.ecn.ac.uk/).

## References


Appleby, P. G. & F. Oldfield, 1978. The calculation of 210Pb dates assuming a constant rate of supply of unsupported 210Pb to the sediment. Catena 5: 1–8.

Appleby, P. G., P. J. Nolan, D. W. Gifford, M. J. Godfrey, F. Oldfield, N. J. Anderson & R. W. Battarbee, 1986. 210Pb dating by low background gamma counting. Hydrobiologia 141: 21–27.

Appleby, P. G., N. Richardson & P. J. Nolan, 1992. Self-absorption corrections for well-type germanium detectors. Nuclear Instruments and Methods in Physics Research B 71: 228–233.

Bailey-Watts, A. E. & A. Kirika, 1987. A re-assessment of the phosphorus inputs to Loch Leven (Kinross, Scotland): rationale and an overview of results on instantaneous loadings with special reference to runoff. Transactions of the Royal Society of Edinburgh, Earth Sciences 78: 351–367.

Bailey-Watts, A. E. & A. Kirika, 1996. A re-assessment of the phosphorus loading to Loch Leven (Kinross, Tayside) – 1995. Report to Scottish Natural Heritage and Scottish Environment Protection Agency: 30 pp.

Bailey-Watts, A. E. & A. Kirika, 1999. Poor water quality in Loch Leven (Scotland) in 1995, in spite of reduced phosphorus loadings since 1985: the influences of catchment management and inter-annual weather variation. Hydrobiologia 403: 135–151.

Bailey-Watts, A. E., R. Sargent, A. Kirika & M. Smith, 1987. Loch Leven phosphorus loading. Report to Department of Agriculture and Fisheries for Scotland, Nature Conservancy Council, Scottish Development Department and Tayside Regional Council: 46 pp.

Battarbee, R. W., V. J. Jones, R. J. Flower, N. G. Cameron, H. Bennion, L. Carvalho & S. Juggins, 2001. Diatoms. In Smol, J. P., H. J. B. Birks & W. M. Last (eds), Tracking Environmental Change Using Lake Sediments. Volume 3: Terrestrial, Algal, and Siliceous Indicators. Kluwer Academic Publishers, Dordrecht: 155–202.

Bennion, H., S. Juggins & N. J. Anderson, 1996. Predicting epilimnetic phosphorus concentrations using an improved diatom-based transfer function and its application to lake eutrophication management. Environmental Science and Technology 30: 2004–2007.

Bennion, H., J. Fluin & G. L. Simpson, 2004. Assessing eutrophication and reference conditions for Scottish freshwater lochs using subfossil diatoms. Journal of Applied Ecology 41: 124–138.

Brett, M. T. & M. M. Benjamin, 2008. A review and reassessment of lake phosphorus retention and the nutrient loading concept. Freshwater Biology 53: 194–211.

Caines, L. A. & R. Harriman, 1976. Chemical investigations. In Loch Leven Annual Report 1976: 1–6.

Carpenter, S. R., N. E. Caraco, D. L. Correll, W. Howarth, A. N. Sharpley & V. H. Smith, 1998. Non-point pollution of surface waters with phophourus and nitrogen. Ecological Applications 8: 559–568.

Carvalho, L., C. Ferguson, I. Gunn, H. Bennion, B. Spears, A. Kirika & L. May, 2011. Water quality of Loch Leven: responses to enrichment, restoration and climate change. Hydrobiologia. doi:10.1007/s10750-011-0923-x.

Carss, D., B. M. Spears, L. Quinn & R. Cooper, 2011. Long-term variations in waterfowl populations in Loch Leven: identifying discontinuities between local and national trends. Hydrobiologia. doi:10.1007/s10750-011-0927-6.

Castle, K., C. A. Frost & D. F. Flint, 1999. The Loch Leven project – buffer strips in practice on a catchment scale. Aspects of Applied Biology 54: 71–78.

Chaichana, R., R. Leah & B. Moss, 2010. Birds as eutrophicating agents: a nutrient budget for a small lake in a protected area. Hydrobiologia 646: 111–121.

Codd, G. A., 1995. Cyanobacterial toxins: occurrence, properties and biological significance. Water Science Technology 32: 149–156.

Coveney, M. F., E. F. Lowe, L. E. Battoe, E. R. Marzolf & R. Conrow, 2005. Response of a eutrophic, shallow subtropical lake to reduced nutrient loading. Freshwater Biology 50: 1718–1730.

D'Arcy, B. J., 1991. Legislation and control of dye-house pollution. Journal of the Society of Dyers and Colourists 107: 387–389.

D'Arcy, B. J., L. May, J. Long, I. R. Fozzard, S. Greig & A. Brachet, 2006. The restoration of Loch Leven, Scotland, UK. Water Science Technology 53: 183–191.

Defew, L. H., 2008. The influence of high-flow events on phosphorus delivery to Loch Leven, Scotland, UK. PhD Thesis, University of Edinburgh: 275 pp.

Dillon, P. J. & F. H. Rigler, 1974. A test of a simple nutrient budget model predicting phosphorus concentration in lake water. Journal of the Fisheries Research Board of Canada 31: 1771–1778.

Dudley, B. & L. May, 2007. Estimating the phosphorus load to waterbodies from septic tanks. Report to the Scottish Environment Protection Agency and Scottish Natural Heritage: 45 pp.

Dudley, B., I. D. M. Gunn, L. Carvalho, I. Proctor, M. T. O'Hare, K. J. Murphy & A. Milligan, 2011. Changes in aquatic macrophyte communities in Loch Leven – evidence of recovery from eutrophication? Hydrobiologia. doi:10.1007/s10750-011-0924-9.

Eisenreich, S. J., R. T. Bannermann & D. E. Armstrong, 1975. A simplified phosphorus analytical technique. Environmental Letters 9: 45–53.

Elliott, J. A. & L. Defew, 2011. Modelling the response of phytoplankton in a shallow lake (Loch Leven, UK) to changes in lake retention time and water temperature. Hydrobiologia. doi:10.1007/s10750-011-0930-y.

European Parliament, 2000. Directive 2000/60/EC of the European Parliament and of the Council of 23 October 2000, establishing a framework for community action in the field of water policy. Official Journal of the European Communities L327(1): 1–72.

Ferguson, R. I., 1986. River loads underestimated by ratings curves. Water Research 22: 74–76.

Frost, C. A., 1994. Soil erosion: Loch Leven. Report to Scottish Natural Heritage: 16 pp.

Frost, A., 1996. Loch Leven and diffuse pollution. In Petchey, A. M., B. J. D'Arcy & C. A. Frost (eds), Diffuse pollution and agriculture. Scottish Agricultural College, Aberdeen: 174–182.

Gulati, R. D. & E. van Donk, 2002. Lakes in the Netherlands, their origin, eutrophication and restoration: state-of-the-art review. Hydrobiologia 478: 73–106.

Gunn, I. D. M., M. T. O'Hare, P. S. Maitland & L. May, 2011. Long-term trends in Loch Leven invertebrate communities. Hydrobiologia.doi:10.1007/s10750-011-0926-7.

Hancock, C.G., 1982. Sources and utilisation of nutrients in the Loch of Strathbeg, Aberdeenshire. PhD Thesis, University of Aberdeen: 308 pp.

Harper, D., 1992. Eutrophication of freshwaters: principles problems and restoration. Chapman & Hall, London: 327.

Holden, A. V. & L. A. Caines, 1974. Nutrient chemistry of Loch Leven. Proceedings of the Royal Society Edinburgh B 74: 101–122.

Holden, A., L. A. Caines & R. Harriman, 1975. Chemical investigations. In Loch Leven Annual Report, 1975: 3–8.

Jeppesen, E., M. Søndergaard, J. P. Jensen, K. Havens, O. Anneville, L. Carvalho, M. F. Coveney, R. Deneke, M. Dokulil, B. Foy, D. Gerdeaux, S. E. Hampton, K. Kangur, J. Köhler, S. Körner, E. Lammens, T. L. Lauridsen, M. Manca, M. Miracle, B. Moss, P. Nõges, G. Persson, G. Phillips, R. Portielje, S. Romo, C. L. Achelske, D. Straile, I. Tatrai, E. Willén & M. Winder, 2005. Lake responses to reduced nutrient loading – an analysis of contemporary long-term data from 35 case studies. Freshwater Biology 50: 1747–1771.

Jeppesen, E., M. Sondergaard, T. L. Lauridsen, B. Kronvang, M. Bekiloglu, E. Lammens, H. S. Jensen, J. Kohler, A. M. Ventela, M. Tarvainen & I. Tatrai, 2007. Danish and other European experiences in managing shallow lakes. Lake and Reservoir Management 23: 439–445.

Juggins, S. 2003. C2 User guide. Software for ecological and palaeoecological data analysis and visualisation. University of Newcastle, Newcastle upon Tyne: 69 pp.

Kirby, R. P., 1971. The bathymetrical resurvey of Loch Leven, Kinross. Geography Journal 137: 372–378.

Kohler, J., S. Hilt, R. Adrian, A. Nicklisch, H. P. Kozerski & N. Walz, 2005. Long-term response of a shallow, moderately flushed lake to reduced external phosphorus and nitrogen loading. Freshwater Biology 50: 1639–1650.

Leigh, C., M. A. Burford, D. T. Roberts & J. W. Udy, 2010. Predicting the vulnerability of reservoirs to poor water quality and cyanobacterial blooms. Water Research 44: 4487–4496.

LLCMP, 1999. Loch Leven Catchment Management Plan: The Report of the Loch Leven Area Management Advisory Group: 93 pp.

Mackereth, F. J. H., 1969. A short core sampler for sub-aqueous deposits. Limnology and Oceanography 14: 145–151.

Manny, B. A., W. C. Johnson & R. G. Wetzel, 1994. Nutrient additions by waterfowl to lakes and reservoirs: Predicting their effects on productivity and water quality. Hydrobiologia 279–280: 121–132.

May, L. & L. Carvalho, 2010. Maximum growing depth of macrophytes in Loch Leven, Scotland, United Kingdom, in relation to historical changes in estimated phosphorus loading. Hydrobiologia 646: 123–131.

May, L. & B. M. Spears, 2011a. A history of scientific research at Loch Leven, Kinross, Scotland. Hydrobiologia. doi: 10.1007/s10750-011-0929-4.

May, L. & B. M. Spears, 2011b. Managing ecosystem services at Loch Leven, Scotland, UK: actions, impacts and unintended consequences. Hydrobiologia. doi:10.1007/s10750-011-0931-x.

Morgan, N. C., 1974. Historical background to International Biological Program Project at Loch Leven, Kinross.

Proceedings of the Royal Society of Edinburgh Section B 74: 45–55.
Munro, D., 1994. Loch Leven and the River Leven – A Landscape Transformed. The River Leven Trust, Markinch: 196.
Murphy, J. & J. P. Riley, 1962. A modified single solution method for the determination of phosphate in natural waters. Analytica Chimica Acta 27: 31–36.
Nürnberg, G. & B. D. LaZerte, 2004. Modeling the effect of development on internal phosphorus load in nutrient poor lakes. Water Resources Research 40: 1–9.
Perth & Kinross Council, 2004. Kinross Area Local Plan 2004: 106 pp.
Pretty, J. N., C. Mason, D. Nedwell, R. Hine, S. Leaf & R. Dils, 2003. Environmental costs of freshwater eutrophication in England and Wales. Environmental Science and Technology 37: 201–208.
Ronicke, H., R. Doerffer, H. Siewers, O. Buttner, K. E. Lindenschmidt, P. Herzsprung, M. Beyer & H. Rupp, 2008. Phosphorus input by nordic geese to the eutrophic Lake Arendsee, Germany. Fundamental and Applied Limnology 172: 111–119.
Sargent, R. J. & D. C. Ledger, 1992. Derivation of a 130 year runoff record for the Loch Leven catchment, south-east Scotland. Proceedings of the Institute of Civil Engineers Water, Maritime and Energy 96: 71–80.
Sas, H., 1989. Lake Restoration by Reduction of Nutrient Loading: Expectations, Experiences, Extrapolations. Acad Verlag Richarz, Berlin.
Schindler, D. W., 1977. Evolution of phosphorus limitation in lakes. Science 195: 260–262.
Schindler, D. W., 2006. Recent advances in the understanding and management of eutrophication. Limnology and Oceanography 51: 356–363.
Smith, I. R., 1974. The structure and physical environment of Loch Leven, Scotland. Proceedings of the Royal Society Edinburgh Section B 74: 82–99.
Sondergaard, M. & E. Jeppesen, 2007. Anthropogenic impacts on lake and stream ecosystems, and approaches to restoration. Journal of Applied Ecology 44: 1089–1094.
Sondergaard, M., P. J. Jensen & E. Jeppesen, 2001. Retention and internal loading of phosphorus in shallow, eutrophic lakes. The Scientific World Journal 1: 427–442.
Søndergaard, M., J. P. Jensen & E. Jeppesen, 2003. Role of sediment and internal loading of phosphorus in shallow lakes. Hydrobiologia 506(509): 135–145.
Søndergaard, M., E. Jeppesen, T. Lauridsen, S. C. H. Van Nes, R. Roijackers, E. Lammens & R. Portielje, 2007. Lake restoration: successes, failures and long-term effects. Journal of Applied Ecology 44: 1095–1105.
Spears, B., L. Carvalho, R. Perkins A. Kirika & D. Paterson, 2011. Long-term variation and regulation of internal loading in Loch Leven. Hydrobiologia. doi:10.1007/s10750-011-0927-6.
Stevens, R. J. & R. V. Smith, 1978. A comparison of discrete and intensive sampling for measuring river loads of nitrogen and phosphorus in the Lough Neagh system. Water Research 11: 631–636.
Vinten, A. J. A., C. Crawford, L. Cole, D. I. McCracken, G. Sym, A. Duncan, & M. N. Aitken, 2004. Evaluating the impact of buffer strips and rural BMPs on water quality and terrestrial biodiversity. In Lewis, D. & L. Gairns (eds), Agriculture and the Environment: Water Framework Directive and Agriculture. Proceedings of the SAC and SEPA Biennial Conference, Edinburgh, 24–25 March 2004: 42–50.
Vollenweider, R. A., 1968. The scientific basis of lake eutrophication, with particular reference to phosphorus and nitrogen as eutrophication factors. Technical Report, Organisation of Economic Cooperation and Development, Paris: 159 pp.
Winfield, I. J., C. E. Adams, J. D. Armstrong, R. Gardiner, A. Kirika, J. Montgomery, B. M. Spears, D. C. Stewart, J. E. Thorpe & W. Wilson, 2011. Changes in the fish community of Loch Leven: untangling anthropogenic pressures. Hydrobiologia. doi:10.1007/s10750-011-0923-8.

Hydrobiologia (2012) 681:23–33
DOI 10.1007/s10750-011-0921-z

LOCH LEVEN RESEARCH

# Long-term variation and regulation of internal phosphorus loading in Loch Leven

**B. M. Spears · L. Carvalho · R. Perkins · A. Kirika · D. M. Paterson**

Published online: 3 November 2011


**Abstract** Long-term monitoring data (1968–2008) were used to investigate internal phosphorus (P) loading following external P loading reduction in shallow Loch Leven, Scotland. A whole-lake sediment P inventory (upper 3 cm of sediment; 2005) suggested a release-potential of 29.7 tonnes (t) from the release sensitive sediment P pools. 18.5 t was contained within shallow water sediments (<4.5 m water depth) with 7.6 t in deeper water sediments below the photic zone (>5 m water depth). The "observed" release (<5.1 t), estimated using a water column P mass balance approach (1989–2008), was <5.1 t, indicating the presence of regulating mechanisms. Observed P release declined between 1989 and 2008, with the exception of 2003–2006. Observed P release estimates were positively correlated with annual average water column P concentration after 1989, highlighting the role of internal loading in maintaining poor water quality conditions after management intervention. Multiple regression analysis suggested that internal loading was driven by the wave mixed depth in spring (positive driver), summer water temperature (positive driver) and spring water clarity transparency (negative driver). The potential importance of biological and physico-chemical feedback mechanisms in the regulation of benthic–pelagic coupling and water quality in Loch Leven are discussed.

**Keywords** Sediment · Phosphorus · Internal loading · Weather · Water quality · Recovery · Eutrophication

Guest editors: L. May & B. M. Spears / Loch Leven: 40 years of scientific research

B. M. Spears (✉) · L. Carvalho · A. Kirika
Centre for Ecology & Hydrology, Penicuik, Midlothian EH26 OQB, UK
e-mail: spear@ceh.ac.uk

R. Perkins
School of Earth, Ocean and Planetary Sciences, Cardiff University, Main Building, Park Place, Cardiff CF10 3YE, UK

D. M. Paterson
Sediment Ecology Research Group, Gatty Marine Laboratory, University of St. Andrews, Fife KY16 8LB, UK

D. M. Paterson
Scottish Oceans Institute, University of St. Andrews, Fife KY16 8LB, UK

## Introduction

Elevated phosphorus (P) inputs to Loch Leven from its catchment over many decades (May et al., 2011) resulted in a deterioration of water quality, including high water column total P (TP) concentrations, increased phytoplankton biomass and, consequently, decreased water clarity (Bailey-Watts & Kirika, 1999; Carvalho et al., 2011), and a reduction in macrophyte biomass (expressed as maximum growing depth; May

& Carvalho, 2010). In an attempt to improve water quality, external P inputs to the loch were reduced by more than 60% between the 1970s and 1990s (Bailey-Watts & Kirika, 1999; May et al., 2011). The lake responded slowly and significant improvements in water quality as a result of this management activity have been observed only recently (D'Arcy et al., 2006). This suggests a recovery period of at least 20 years at this site (1989–2008; Fig. 1).

It is generally accepted that this prolonged recovery is driven by re-equilibration processes ("internal loading"), whereby sediment P that has been accumulated over periods of high external loading is released to the overlying water column during the period of recovery in shallow lakes (Sas, 1989). The magnitude of internal loading may diminish slowly as a result of P relinquishment from the system or may be disrupted (capped) relatively quickly through the establishment of natural buffering systems at the sediment–water interface (e.g. macrophytes and benthic algae; Scheffer, 2001; Mehner et al., 2008).

It is hypothesised that a switch in water column nutrient source dependency from the catchment to the sediment has occurred in Loch Leven. This phenomenon is commonly observed in shallow lakes following the reduction of external P loading (Sas, 1989; Jeppesen et al., 2005; Søndergaard et al., 2005). However, although attempts have been made to investigate the importance of internal loading in Loch Leven, through sediment P inventories and water column mass balance estimates, little is known about changes in the relative importance of internal versus external loading across the well documented history of eutrophication and subsequent management and recovery at this site.

Holden & Caines (1974) acknowledged the potential importance of sediment P sources in driving water quality at Loch Leven between 1964 and 1971. They estimated that the mass of dormant release sensitive P (i.e. that associated with Fe and, therefore, releasable under reducing conditions) in the upper 4 cm of sediment was about 20 t P. Using a water column mass balance approach they also estimated that P cycling between the sediments and the water column amounted to a maximum sediment uptake of 136 kg TP day$^{-1}$, a maximum sediment release of 88 kg TP day$^{-1}$. Advances in chemical speciation techniques facilitated a more comprehensive assessment of sediment P fractions in 1990, immediately

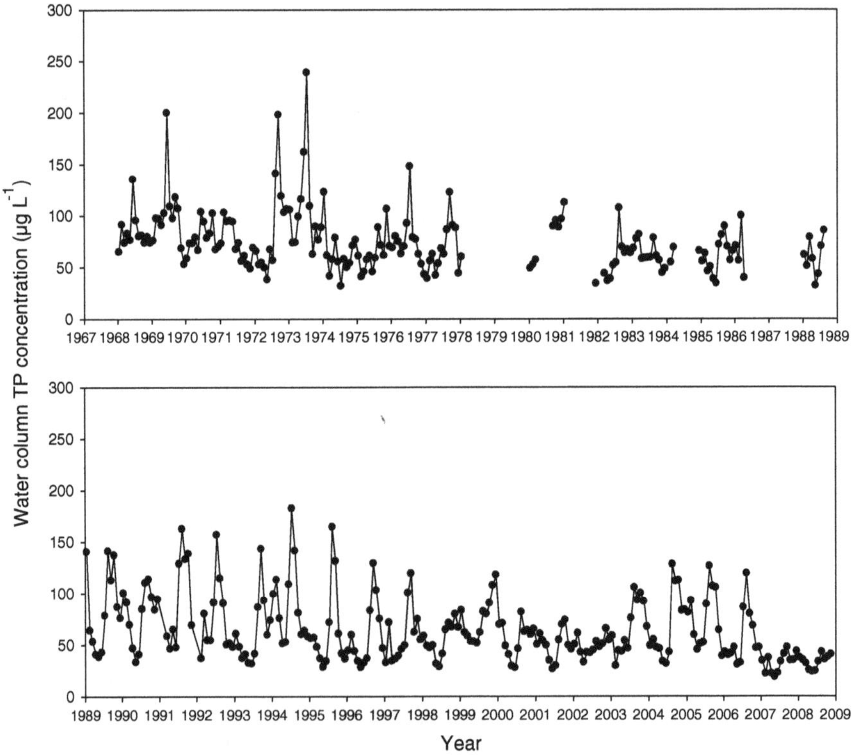


**Fig. 1** Monthly average TP concentrations in the water column of Loch Leven, 1968–2008

after significant P loading reductions (Farmer et al., 1994). The authors found that a large proportion (~30%) of the 68.8 t of TP contained in the upper 4 cm of sediments within the loch was release sensitive. A more recent study (2004–2005; i.e. during the recovery period) found significant spatial (with overlying water depth) and seasonal variations in P composition and concentrations within the active layer of surface sediment (upper 3 cm; Spears et al., 2006, 2007). This work clearly highlighted the need to include annual and spatial averages in constructing representative sediment P inventories.

Characteristic water column TP peaks in late summer/early autumn, which are typically driven by internal loading, have been recorded in Loch Leven since 1989 (Fig. 1, lower panel; Spears et al., 2006, 2008). Such peaks are commonly observed in "recovering" shallow lakes (Sas, 1989; Søndergaard, 2007) at this time of year when external nutrient inputs are low. At Loch Leven, there is also a trough in the hydrograph at this time of year (Smith, 1974). For these reasons, changes in water column TP concentration over this period can be attributed fairly reliably to internal loading.

Recent studies have highlighted strong spatial zonation in relation to the environmental drivers of sediment P release in Loch Leven. Dominant nutrient cycling mechanisms are synthesised from Spears et al. (2006, 2007, 2009), the spatial distribution of dominant autotrophic groups is summarised from Spears (2007) and Spears et al. (2009) and the substrate composition and light attenuation are summarised from Calvert (1974) and unpublished data (2004–2005), respectively. In shallow, photic zones, P release is expected to be driven by wind induced disturbance of sediments and diffusive release across an aerobic sediment–water interface (Fig. 2; Table 1). In these zones, sediment oxygen concentrations are likely maintained through a combination of wind induced wave mixing (Spears & Jones, 2010) and benthic primary production (Spears et al., 2008). Sediment P release in deeper, relatively undisturbed, aphotic sediments is likely dominated by diffusive release across an aerobic (especially in cooler, windy winter/spring) or anoxic (especially in warm, calm summer/autumn) sediment–water interface (Spears et al., 2007). Although the relative importance of these processes is known to vary seasonally and inter-annually, the combined effects of these drivers have not been quantified.

In Loch Leven, wave mixing of sediments and, therefore, the threat of disturbance driven P-release is lowest in summer and highest in winter (Spears & Jones, 2010). In contrast, temperature driven release, through enhanced diffusion rates (shallow and deep water sediments) or the onset of anoxia and resultant release of reductant-soluble P (more likely in deep water sediments), is more important in summer and autumn (Spears et al., 2007, 2008). The effects of benthic autotrophs in reducing sediment P release are likely to be strongest during summers with high temperatures (Spears et al., 2008) that follow an intense spring clear water phase.

The objectives of this article are (1) to use sediment P concentration data to create a whole loch sediment P inventory and partition this spatially according to overlying water depth, (2) to use water column TP fluctuations, associated with apparent internal loading processes, to estimate long-term variations in the magnitude of apparent internal P loading, and (3) to investigate the relative seasonal effects of temperature, wind induced wave mixing, and water transparency (Secchi depth) on long-term variations in internal loading.

## Methods

### Study site

Loch Leven (Fig. 3) is a large (13.3 $km^2$ surface area) shallow (mean depth 3.9 m) eutrophic loch in east central Scotland (latitude 56°10′N, longitude 3°30′W). The loch has a long and well documented history of eutrophication problems, with catchment management during 1970s–1990s resulting in a significant reduction in external P loading (Bailey-Watts & Kirika, 1999; D'Arcy et al., 2006; May et al., 2011). The period of high external loading (1968–1989) was characterised by high baseline TP concentrations apparently masking any seasonal TP trends (Fig. 1). The external TP load was halved between 1985 (20 t TP per year) and 1995 (~8 t TP per year) and appears to have changed little between 1995 and 2005 (8 t TP per year) (May et al., 2011). About 65% of this reduction was conducted between 1970 and 1987 the remaining 25% occurring between 1993 and 1997 (LLAMAG, 1993). A change in the TP signal is evident post 1989 where autumnal internal loading peaks are observed coupled

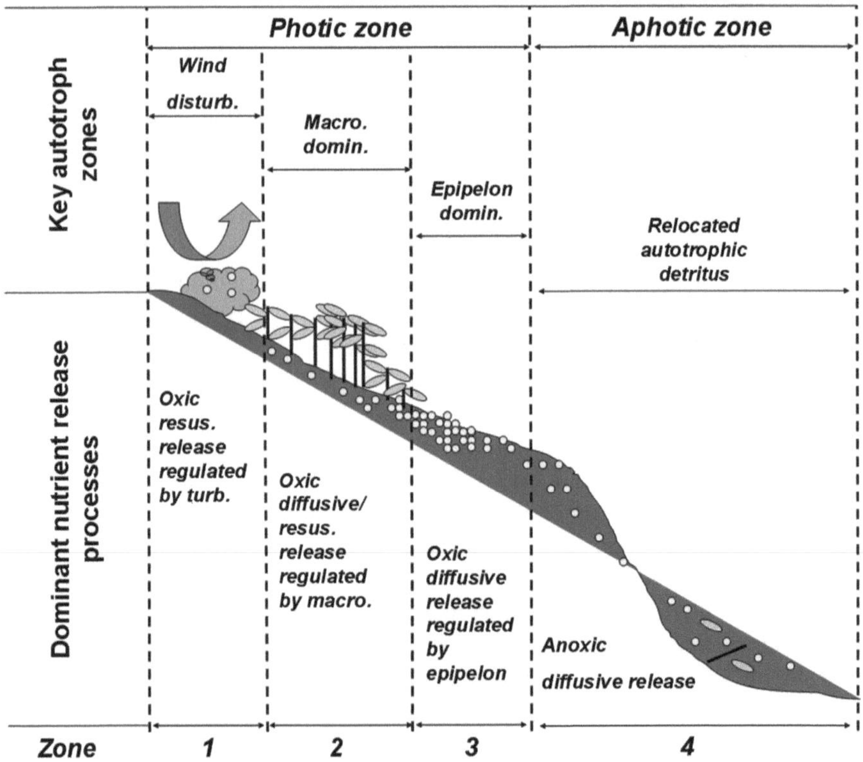


**Fig. 2** Conceptual diagram of dominant zones of sediment P release regulation across Loch Leven. *Disturb.* disturbance, *Resus.* resuspension, *turb.* turbulence

**Table 1** Estimated zonation of functional autrotrophs in Loch Leven based on data from 2004 to 2006; depth ranges represent zones of dominance based on spatial assessments of biomass

| Zone | Dominant autotrophic group | Dominant substratum | Depth (m) | Loch surface area (%) | Incident light reaching sediment surface (%) | Nutrient cycling mechanism |
|---|---|---|---|---|---|---|
| 1 | Microalgae/emergent vegetation | Sand/stones | <1.0 | 21 | 100.0–16.1 | Aerated, diffuse, disturbance |
| 2 | Macrophytes | Sand/mud | 1.0–3.5 | 37 | 16.0–1.4 | Aerated, diffuse, disturbance |
| 3 | Microalgae | Mud | 3.6–5.5 | 23 | 1.3–0.1 | Aerated diffuse, undisturbed |
| 4 | Detritus | Mud | >5.5 | 19 | 0.1–0.0 | Aerated/anoxic, undisturbed |

with a drop in baseline TP. The magnitude of this peak is observed to fluctuate and is lowest during 2000–2003 and 2007–2008.

The substrate of Loch Leven is mainly sand and is dominated mainly by cobbles at depths of less than 2 m, and by silt and organic mud at depths of greater than 2 m (Calvert, 1974; Spears et al., 2006).

## Estimating the sediment P inventory

Sediment P concentrations across a gradient of 1.5–22.5 m overlying water depth (Reed Bower site also sampled; 3.5 m water depth) were reported by Spears et al. (2007). Surface sediments were sampled using a Jenkin surface sediment sampler. Sediment P pools were quantified using the modified Psenner extraction scheme (Farmer et al., 1994). Annual average concentrations of sediment P pools (labile, reductant-soluble and total sediment P) were calculated for the upper 3 cm of surface sediments, using monthly concentrations between April 2004 and April 2005, and extrapolated over a range of water depths (Table 2). Labile and reductant-soluble P were combined as release sensitive sediment P. These annual

**Fig. 3** Bathymetric map of Loch Leven showing the location of the Reed Bower sampling site

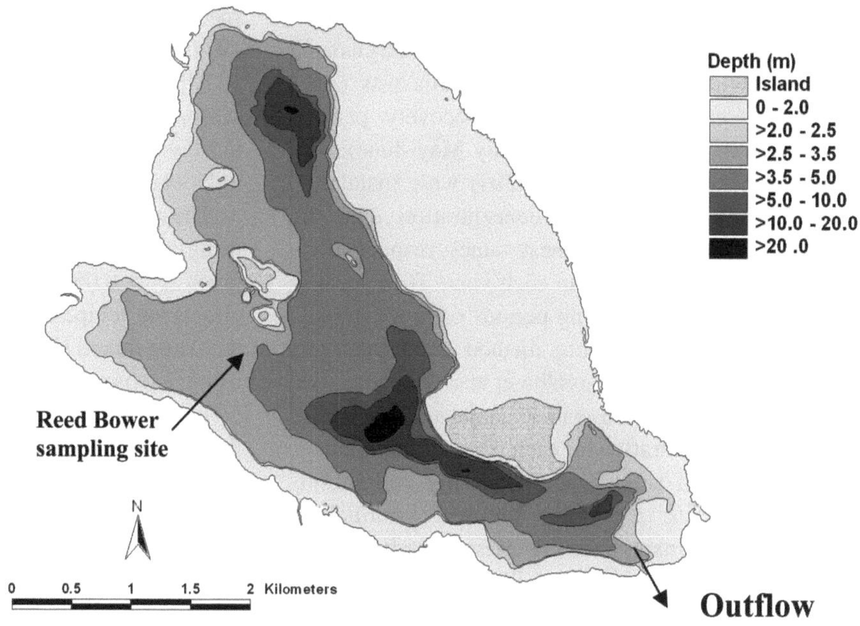


**Table 2** Estimates of sediment surface area, TP content of upper 3 cm of sediment, and release sensitive (labile and reductant-soluble sediment P pools) P for given depth ranges in Loch Leven 2005

| Depth range (m) | Sediment area | | TP content | | Release sensitive P | |
|---|---|---|---|---|---|---|
| | (km$^2$) | (% lake area) | (tonnes) | (% Total) | (tonnes) | (% Total) |
| 2.0–4.4 | 5.44 | 39.6 | 62.5 | 67.9 | 18.6 | 20.1 |
| 4.5–6.9 | 1.87 | 13.6 | 18.8 | 20.4 | 7.0 | 7.6 |
| 7.0–9.4 | 0.37 | 2.7 | 3.3 | 3.4 | 1.4 | 1.5 |
| 9.5–11.9 | 0.23 | 1.7 | 2.1 | 2.3 | 0.8 | 0.9 |
| 12.0–14.4 | 0.16 | 1.1 | 1.4 | 1.5 | 0.5 | 0.5 |
| 14.5–16.9 | 0.17 | 1.2 | 1.5 | 1.6 | 0.5 | 0.6 |
| 17.0–19.4 | 0.11 | 0.8 | 0.9 | 1.0 | 0.4 | 0.4 |
| 19.5–21.9 | 0.09 | 0.7 | 0.8 | 0.9 | 0.3 | 0.3 |
| 22.0–25.0 | 0.07 | 0.5 | 0.5 | 0.6 | 0.2 | 0.2 |
| Total | 8.49 | 61.9 | 92.0 | | 29.7 | 32.3 |

average concentrations were converted to mass of P (tonnes) by correcting for the annual average dry weight of sediment and the surface area of the loch bed for each depth class (Kirby, 1971).

## Estimating long-term variations in internal loading

Water samples for TP analysis were collected at a two-week interval from the Reed Bower sample site (Fig. 3) using an integrated water column sampler deployed to about 25 cm above the sediment surface. Only values for 1989–2008 were used in the analyses that follow. When open water sampling was deemed to be impractical (e.g. due to ice cover), surface water samples were collected from the outflow. The method used for TP analysis was as described by Wetzel & Likens (2000), with an added acidification step (0.1 ml of 30% $H_2SO_4$ was added to the samples before addition of persulfate).

Internal loading was estimated using a water column mass balance approach. The magnitude of the May–September internal loading peak (I-Load$_{M\text{-}S}$; Fig. 1, lower panel; Spears et al., 2006, 2008) was corrected for loch volume to estimate the mass of P release (tonnes) required to effect the observed change in water column TP concentration. This was conducted for each year during the recovery period (1989–2008). The troughs (generally May–June) and peaks (generally August–September) were manually selected from the raw TP concentration data to represent the lowest and highest values, respectively, corresponding to the peak in each year. These peaks occur in the summer, during periods of low external loading and, as such, this method represents an estimate of I-Load$_{M\text{-}S}$ (Carvalho et al., 1995).

I-Load$_{M\text{-}S}$ was plotted against annual average water column TP concentration for each year, the covariance between the two being quantified using correlation analysis ($n = 19$; $\alpha = 0.05$) following normality criteria testing using the Anderson–Darling test ($P > 0.05$; Townend, 2004). All statistical analyses were conducted using Minitab version 15.

## Identifying the long-term drivers of internal loading

The effects of water clarity (Secchi depth), wind induced wave mixed depth, and temperature on the magnitude of I-Load$_{M\text{-}S}$ were assessed between 1989 and 2007, excluding 2000-2002 due to lack of wind data ($n = 16$). Secchi depth was recorded at the Reed Bower sampling site (Fig. 3) at roughly two-weekly intervals. Monthly mean temperatures were calculated from daily measurements recorded at a weather station on the shore by Kinross Estates. Average daily temperatures were calculated as the arithmetic mean of the maximum and minimum daily temperatures recorded. Gaps in the on-shore weather data were filled using a regression equation that related 30 years of air temperature records from Loch Leven ($T_{\mathrm{Leven}}$) to those recorded at Leuchars ($T_{\mathrm{Leuchars}}$), a weather station situated about 44 km north-east of the loch (http://www.metoffice.gov.uk/climate/uk/stationdata/). This was as follows:

$$T_{\mathrm{Leven}} = 1.023 \times T_{\mathrm{Leuchars}} - 0.9938(r^2 = 0.9885)$$

Daily average wind speed and direction for Loch Leven was provided by Kinross Estates. Daily wave mixed depths, averaged over six sampling locations across the loch (see Spears & Jones, 2010) were estimated as a function of effective fetch, and wind speed and direction, using simple wave theory (Smith, 1974; Douglas & Rippey, 2000).

All data were averaged, first by month and then by season (winter: December to February; spring: March to May; summer: June to August; autumn: September to November).

The normality of each data population was assessed using an Anderson–Darling test and assumed normal given $P > 0.05$ (Townend, 2004). The combined effects of temperature, Secchi depth, and depth of the wave mixed layer on I-Load$_{M\text{-}S}$ were investigated using stepwise multiple regression analysis that included seasonal and annual averages of each of the three drivers. Planned multiple regression was conducted using the output of the stepwise multiple regression and the resultant model was validated against estimates of I-Load$_{M\text{-}S}$.

# Results

## Sediment P inventory

The depth range of sediment studied (2.0–25.0 m) represents about 62% of the surface area of the loch (Table 2). Depth range 2.0–4.4 m (40% loch surface area) had the highest surface sediment TP (62.5 t; 68% of whole loch sediment TP estimate) and release sensitive P mass (18.6 t; 20% of whole loch sediment TP estimate). This was followed by the 4.5–6.9 m depth range which contained 19 t of sediment TP (20% whole loch sediment TP) and 7 t of release sensitive P (8% whole loch sediment TP). Only 11 t sediment TP (11% whole loch sediment TP) and 4 t release sensitive P (4% whole loch sediment TP) was contained within the depth range 7.0–25.0 m (9% loch surface area).

## Long-term variation in internal loading

During the recovery period, internal loading (I-Load$_{M\text{-}S}$) was highest (5.0 t TP) in 1991 and lowest in 2002 (1.1 t TP) (Fig. 4). However, overall, there was a steadily declining trend in I-Load$_{M\text{-}S}$ from 1989 to 2002. This was followed by a temporary increase in values between 2003 and 2006, followed by a large decrease observed in 2007 to a value that was sustained in 2008. Annual average water column TP concentrations were significantly and positively correlated with I-Load$_{M\text{-}S}$ (correlation coefficient = 0.794; $P = < 0.001$).

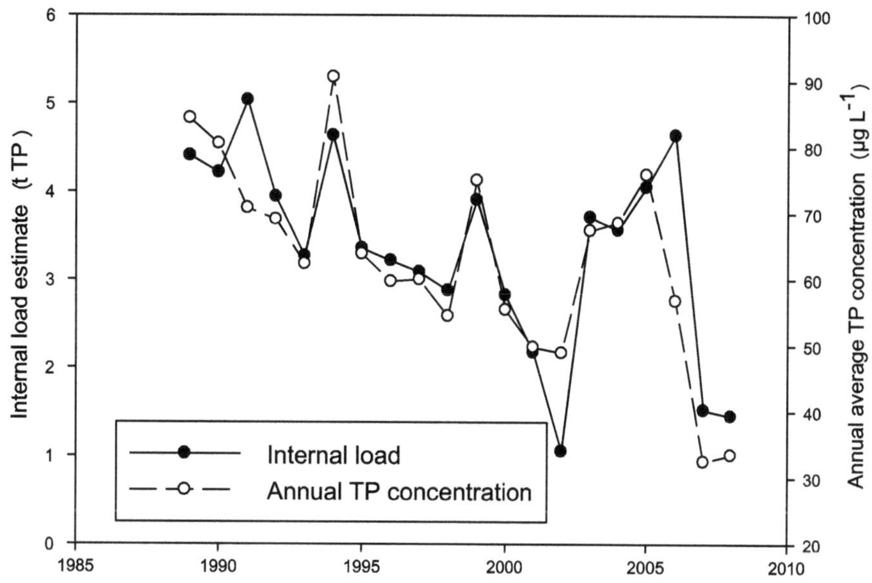


**Fig. 4** Internal load estimates and annual mean TP concentration for Loch Leven, 1989–2008

### Identifying the drivers of internal P loading

The effects of temperature, Secchi depth, and wave mixing on internal loading were first assessed using stepwise multiple regression analysis. A significant result was returned in which the spring wave mixed depth had the largest effect, followed by summer air temperature and then by spring Secchi depth (Table 3). Standard multiple regression analysis was then used to construct the following I-Load$_{M\text{-}S}$ model:

$$\text{I-Load}_{M-S}(\text{t TP}) = -5.86 + 2.03\text{Zc} + 0.53T - 0.75\ \text{SD}$$

where I-Load$_{M\text{-}S}$ is the Internal P load (t TP), Zc is the wave mixed depth (m), $T$ is the summer air temperature (°C), SD is the Secchi depth (m).

The fit of the modelled data to the measured data is shown (Fig. 5). There was a significant correlation (correlation coefficient = 0.84; $P < 0.001$) between modelled and "observed" I-Load$_{M\text{-}S}$.

## Discussion

### Distribution of sediment P in relation to the regulation of release

The difference between the mass of release sensitive P in the sediments (total 29.7 t P) and the mass of P required to raise the water column P concentration during the internal loading peak in Loch Leven in 2005 (~4 t TP) is large. This suggests that only a small

**Table 3** Results of the multiple linear regression analysis of drivers of internal load at Loch Leven

| Predictor | | Coefficient | SE coefficient | $T$ | $P$ |
|---|---|---|---|---|---|
| Constant | | −5.86 | 3.3 | −1.8 | 0.11 |
| Spring wave mixed depth | | 2.03 | 0.7 | 2.9 | 0.02 |
| Summer air temperature | | 0.52 | 0.2 | 2.8 | 0.02 |
| Spring Secchi depth | | −0.75 | 0.5 | −1.6 | 0.15 |
| **Analysis of variance** | **DF** | **SS** | **MS** | **$F$** | **$P$** |
| Regression | 3 | 6.3 | 2.1 | 8.39 | 0.006 |
| Residual error | 9 | 2.2 | 0.2 | | |
| Total | 12 | 8.5 | | | |
| | $S = 0.49$ | | $R^2 = 0.74$ | | $R^2$ (adj.) = 0.65 |

*SE* standard error, *DF* degrees of freedom, *SS* sum of squares, *MS* mean square

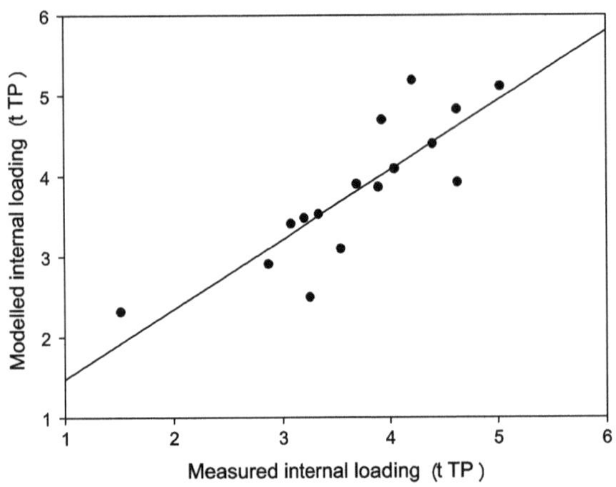


**Fig. 5** Relationship between modelled internal load and measured internal load for Loch Leven, 1989–2008 ($r^2 = 0.84$; $P < 0.001$)

amount of the sediment P pool is responsible for the observed changes in water column TP during internal loading events at this site. This situation is often observed in shallow lakes and can be explained by considering spatial variation in lake bed characteristics in relation to the available P stock (Boström et al., 1988).

The spatial distribution of sediment P pools in Loch Leven was assessed in 2004/2005 (Spears et al., 2007, 2008). These studies highlighted co-variation between reductant-soluble sediment P concentrations, bottom water dissolved oxygen concentrations and benthic algal biomass, patterns that are consistent with observations from other lakes (van Luijn et al., 1995; Jin et al., 2006). Macrophytes have also been reported to reduce sediment P release through similar mechanisms in lakes (Granéli & Solander, 1998), although this has not been directly quantified in Loch Leven. In 2005, sediment P release may have been buffered by benthic autotrophs in sediments at a depth of between 1.0 and 5.5 m in Loch Leven; the lower boundary being regulated by light attenuation. It follows that sediment P release will have been unimpeded by benthic autotrophs in regions of the loch bed that lie more than 5.5 m below the water surface. Further, deeper water sediments in Loch Leven were susceptible to net-heterotrophy resulting in reduced dissolved oxygen concentrations in bottom waters and the reduction of redox-sensitive P complexes associated with an increase in sediment P release (Spears et al., 2007). This was significantly enhanced in the warmer summer months (Spears et al., 2007).

In 2005, 7 t of release sensitive P was estimated to be contained within surface sediments in the aphotic zone (i.e. >5.5 m overlying water depth). This compares with some 23 t of release sensitive P in the photic zone (i.e. <5.5 m overlying water depth). Due to strong concentration gradients in dissolved P across the sediment water interface (Spears et al., 2007), it is unlikely that diffusive P release from aerobic sediments in the photic zone is negligible (Jiang et al., 2006; Spears et al., 2008). However, release mechanisms such as wind induced wave disturbance (Søndergaard et al., 1992) and bioturbation (Andersson et al., 1988) are also likely to be important in these well aerated zones. In contrast, the dominant release mechanisms in the aphotic zone are more likely to be driven by temperature related microbial deoxygenation of surface sediments and the associated release of redox-sensitive P complexes (Boström et al., 1988; Spears et al., 2007).

## Understanding the long-term regulation of internal loading

Under the present external P loading regime and photic depth, the magnitude of internal P loading in Loch Leven is driven mainly by variations in weather. Specifically, internal loading appears to increase as a result of high wind and turbidity in spring and high temperature in summer. The direct relationship between internal loading and summer temperature is well documented and results from direct increase in solute mobility across concentration gradients coupled with the metabolic scrubbing of dissolved oxygen by benthic heterotrophs and associated reduction of redox-sensitive sediment P complexes (Søndergaard, 1989; Jenson & Andersen, 1992; Spears et al., 2008). The spring Secchi depth reflects the light conditions at the sediment surface during the period of benthic autotroph establishment and is probably a good proxy for summer benthic autotrophic biomass (Canfield et al., 1985). Phillips et al. (2005) linked improvements in spring water clarity with increased summer benthic algal biomass and reduced summer internal loading. The mechanisms behind spring wave mixing with respect to the regulation of the magnitude of internal load are less easily explained. Wave mixing in spring may result in sediment disturbance which can enhance sediment nutrient release (Søndergaard et al., 1992) and inhibit the establishment of benthic autotrophs (macrophytes: Jupp & Spence, 1977, Doyle, 2001; microalgae: Cyr, 1998a). The negative effects of wave action on the biomass of the macrophytes *Potamogeton filiformus* and *P. Pectinatus* in Loch Leven have been well documented, with up to 80% of the biomass reduction in these species being attributed to wave action and the associated deterioration of substrate towards a coarser nutrient poor material (Jupp & Spence, 1977). This suggests that the buffering capacity of benthic autotrophs may be reduced under high wind conditions in Loch Leven. Additionally, high wave induced mixing has been linked to the focusing of sediments from shallow waters to deeper waters (Hilton, 1985; Cyr, 1998b) and may represent the relocation of P from the photic to the aphotic zone, thus increasing overall sediment P release.

Other studies have highlighted the important role of benthic autotrophs, macrophytes in particular, as drivers of both positive and negative feedbacks through which "recovering" shallow lakes can alternate between clear water macrophyte and turbid phytoplankton-dominated states (Rip et al., 2005; van Nes et al., 2007). The switch between positive and negative feedback mechanisms has also been linked to climate change (i.e. the North Atlantic Oscillation). These effects are many and are linked mainly through direct relationships between indices of the NAO (calculated by comparing sea level atmospheric pressures in the northern (Southwest Iceland) and southern (Gibraltar) reach of the northern hemisphere (Jones et al., 1997)) and precipitation, wind and temperature (Straile et al., 2003). Effects can include variations in nutrient loading from the catchment as a result of high precipitation (Monteith et al., 2000; Rip, 2007), variations in the wave mixed zone and potential sediment disturbance (Loch Leven: Spears & Jones, 2010), and variations in zooplankton grazing rates associated with the intensity of the spring clear water phase (Scheffer et al., 2001).

The low internal loading events of 2007 and 2008 in Loch Leven were characterised by low spring wind speeds, low summer temperatures and high Secchi depths (i.e. high NAO winter index years; http://www.cru.uea.ac.uk/~timo/projpages/nao_update.htm). In addition, a thorough assessment of macrophyte colonisation conducted in 2008 returned a maximum macrophyte colonisation depth similar to that observed before eutrophication (4.6 m: May & Carvalho, 2010), presumably resulting in higher P uptake by both macrophytes and epiphytes (Carpenter & Lodge, 1986). However, these years were also years of high rainfall and presumably, higher flushing rates (Carvalho et al., 2011). Although the results of this study highlight the importance of climate change as a driver of internal loading, it should be noted that interactions between climate and land use drivers may override the observed relationships. Further work is required to quantify the interactions between climate change, catchment pressures and biological control of internal nutrient processes in Loch Leven.

## Conclusions

Improvements in water column TP concentration during the recovery period appear to have resulted from two different processes. First, a gradual decline in TP was observed over the 20 year period (Carvalho et al., 2011). Second, a more instantaneous decline in TP was observed in 2007. Correlation analysis suggests that one important driver of this variation in Loch Leven between 1989 and 2008 was the magnitude of internal loading. Variability in the magnitude of internal loading during the recovery period appeared to be regulated mainly by climate change variables (i.e. wind and temperature) and, to a lesser extent, indicators of water clarity (i.e. Secchi depth), rather than TP concentrations in the sediment.

**Acknowledgments** We would like to acknowledge Mr. Jamie Montgomery and, his father, Sir David Montgomery for their continued support, enthusiasm and assistance throughout the project. We would also like to thank Willie Wilson, of Loch Leven Fisheries, for his invaluable advice and support during the project and for the provision of weather data. Finally, we wish to thank Dr. Martin Søndergaard (NERI, Denmark) and Dr. Andrea Kelly (the Broads Authority, UK) for useful comments that led to the improvement of the manuscript. This project was funded by the Natural Environment Research Council, UK.

## References

Andersson, G., W. Granéli & J. Stenson, 1988. The influence of animals on phosphorus cycling in lake ecosystems. Hydrobiologia 170: 267–284.

Bailey-Watts, A. E. & A. Kirika, 1999. Poor water quality in Loch Leven (Scotland) in 1995, in spite of reduced phosphorus loading since 1985: the influences of catchment management and inter-annual weather variation. Hydrobiologia 403: 135–151.

Boström, B., J. M. Anderson, S. Fleischer & M. Jansson, 1988. Exchange of phosphorus across the sediment–water interface. Hydrobiologia 170: 229–244.

Calvert, S. E., 1974. The distribution of bottom sediments in Loch Leven, Kinross. Proceedings of the Royal Society of Edinburgh B 74: 69–80.

Canfield, D. E. Jr., K. A. Langeland, S. B. Linda & W. T. Haller, 1985. Relations between water transparency and maximum depth of macrophyte colonization in lakes. Journal of Aquatic Plant Management 23: 25–28.

Carpenter, S. R. & D. M. Lodge, 1986. Effects of submersed macrophytes on ecosystem processes. Aquatic Botany 26: 341–370.

Carvalho, L., M. Beklioglu & B. Moss, 1995. Changes in a deep lake following sewage diversion – a challenge to the orthodoxy of external phosphorus control as a restoration strategy? Freshwater Biology 34: 399–410.

Carvalho, L., C. Ferguson, I. Gunn, H. Bennion, B. Spears, A. Kirika & L. May, 2011. Water quality of Loch Leven: responses to enrichment, restoration and climate change. Hydrobiologia. doi:10.1007/s10750-011-0923-x.

Cyr, H., 1998a. How does vertical distribution of chlorophyll vary in littoral sediments of small lakes? Freshwater Biology 40: 25–40.

Cyr, H., 1998b. Effects of wave disturbance and substrate slope on sediment characteristics in the littoral zone of small lakes. Canadian Journal of Fisheries and Aquatic Sciences 55: 967–976.

D'Arcy, B. J., L. May, J. Long, I. R. Fozzard, S. Greig & A. Brachet, 2006. The restoration of Loch Leven, Scotland. Water Science and Technology 53: 181–191.

Douglas, R. W. & B. Rippey, 2000. The random redistribution of sediment by wind in a lake. Limnology and Oceanography 45: 686–694.

Doyle, R. D., 2001. Effects of waves on the early growth of *Vallisneria americanna*. Freshwater Biology 46: 389–397.

Farmer, J. G., A. E. Bailey-Watts, A. Kirika & C. Scott, 1994. Phosphorus fractionation and mobility in Loch Leven sediments. Aquatic Conservation: Marine and Freshwater Ecosystems 4: 45–56.

Granéli, W. & D. Solander, 1998. Influence of aquatic macrophytes in lakes. Hydrobiologia 170: 245–266.

Hilton, J., 1985. A conceptual framework for predicting the occurrence of sediment focussing and sediment redistribution in small lakes. Limnology and Oceanography 30: 1131–1143.

Holden, A. V. & L. A. Caines, 1974. Nutrient chemistry of Loch Leven, Kinross. Proceedings of the Royal Society of Edinburgh B 74: 101–121.

Jenson, H. S. & F. Ø. Andersen, 1992. Importance of temperature, nitrate, and pH for phosphate release from aerobic sediments of four shallow, eutrophic lakes. Limnology and Oceanography 37: 577–589.

Jeppesen, E., M. Søndergaard, J. P. Jensen, K. Havens, O. Anneville, L. Carvalho, M. F. Coveney, R. Deneke, M. T. Dokulil, B. Foy, D. Gerdeaux, S. E. Hampton, S. Hilt, K. Kangur, J. Kohler, E. H. H. R. Lammens, T. L. Lauridsen, M. Manca, M. R. Miracle, B. Moss, P. Noges, G. Persson, G. Phillips, R. Portielje, S. Romo, C. L. Schelske, D. Straile, I. Tatrai, E. Willen & M. Winder, 2005. Lake responses to reduced nutrient loading – an analysis of contemporary long-term data from 35 case studies. Freshwater Biology 50: 1747–1771.

Jiang, X., X. Jin, Y. Yao, L. Li & F. Wu, 2006. Effects of oxygen on the release and distribution of phosphorus in the sediments under the light condition. Environmental Pollution 141: 482–487.

Jin, X., X. Jiang, Y. Yao, L. Li & F. Wu, 2006. Effects of light and oxygen on the uptake and distribution of phosphorus at the sediment–water interface. Science of the total Environment 357: 231–236.

Jones, P. D., T. Jonsson & D. Wheeler, 1997. Extension to the North Atlantic Oscillation using early instrumental pressure observations from Gibraltar and south-west Iceland. International Journal of Climatology 17: 1433–1450.

Jupp, B. P. & D. H. N. Spence, 1977. Limitations on macrophytes in a eutrophic lake, Loch Leven. II Wave action, sediments, and water fowl grazing. The Journal of Ecology 65: 175–186.

Kirby, R. P., 1971. The bathymetrical resurvey of Loch Leven, Kinross. Geographical Journal 137: 372–378.

LLAMAG, 1993. Loch Leven: the report of the Loch Leven area management advisory group. Forth River Purification Board, Edinburgh.

May, L. & L. Carvalho, 2010. Maximum growing depth of macrophytes in Loch Leven, Scotland, United Kingdom in relation to historical changes in estimated phosphorus loading. Hydrobiologia 646: 123–131.

May, L., L. H. Defew, H. Bennion & A. Kirika, 2011. Historical changes(1905–2005) in external phosphorus loads to Loch Leven, Scotland, UK. Hydrobiologia. doi:10.1007/s10750-011-0922-y.

Mehner, T. M., T. Diekmann, P. Gonsiorczyk, R. Kasprzak, L. Koschel, M. Krienitz, M. Rumpf, G. Schulz & Wauer, 2008. Rapid recovery from eutrophication of a stratified lake by disruption of internal nutrient load. Ecosystems 11: 1142–1156.

Monteith, D. T., C. D. Evans & B. Reynolds, 2000. Are temporal variations in the nitrate content of UK upland freshwaters linked to the North Atlantic Oscillation? Hydrological Processes 14: 1745–1749.

Phillips, G., A. Kelly, J. Pitt, R. Sanderson & E. Taylor, 2005. The recovery of a very shallow lake, 20 years after the control of effluent driver phosphorus. Freshwater Biology 50: 1628–1638.

Rip, W., 2007. Cyclic state shifts in a restored shallow lake. Thesis, Wageningen University. ISBN 978-90-8504-706-3.

Rip, W., M. R. L. Ouboter, E. H. van Ness & B. Beltman, 2005. Long term oscillation of a shallow lake ecosystem upon reduction in external phosphorus load. Archiv für Hydrobiologie 164: 387–409.

Scheffer, M., 2001. Ecology of shallow lakes. Kluwer Academic Press, London.

Scheffer, M., D. Straile, E. H. van Nes & H. Hosper, 2001. Climatic warming causes regime shifts in lake food webs. Limnology and Oceanography 46: 1780–1783.

Sas, H., 1989. Lake restoration by reduction of nutrient loading. Academic Verlag Richarz, St. Augustin.

Smith, I. R., 1974. The structure and physical environment of Loch Leven, Scotland. Proceedings of the Royal Society of Edinburgh B 74: 81–100.
Søndergaard, M., 1989. Phosphorus release from a hypertrophic lake sediment: experiments with intact sediment cores in a continuous flow system. Archive für Hydrobiologie 116: 45–59.
Søndergaard, M., 2007. Nutrient dynamics in lakes – with emphasis on phosphorus, sediment and lake restorations. DSc. Thesis. National Environmental Research Institute, University of Aarhus, Denmark: 276 pp.
Søndergaard, M., P. Kristensen & E. Jeppesen, 1992. Phosphorus release from resuspended sediment in the shallow and wind-exposed Lake Arresø, Denmark. Hydrobiologia 228: 91–99.
Søndergaard, M., J. P. Jensen & E. Jeppesen, 2005. Seasonal response of nutrients to reduced phosphorus loading in 12 Danish lakes. Freshwater Biology 50: 1605–1615.
Spears, B. M., 2007. Benthic–pelagic nutrient cycling in shallow lakes: investigating the functional role of benthic microalgae. PhD Thesis, University of St Andrews.
Spears, B. M. & I. Jones, 2010. The long-term (1978–2005) effects of the North Atlantic oscillation on wind-induced wave mixing in Loch Leven (Scotland). Hydrobiologia 646: 49–59.
Spears, B. M., L. Carvalho, R. Perkins, A. Kirika & D. M. Paterson, 2006. Spatial and historical variation in sediment phosphorus fractions and mobility in a large shallow lake. Water Research 40: 383–391.
Spears, B. M., L. Carvalho, R. Perkins, A. Kirika & D. M. Paterson, 2007. Sediment P cycling in a large shallow lake: spatio-temporal variation in P pools and release. Hydrobiologia 584: 37–48.
Spears, B. M., L. Carvalho, R. Perkins & D. M. Paterson, 2008. The effects of light on sediment nutrient flux and water column nutrient stoichiometry in a shallow lake. Water Research 42: 977–986.
Spears, B. M., I. D. M. Gunn, L. Carvalho, I. J. Winfield, B. Dudley, K. Murphy & L. May, 2009. An evaluation of methods for sampling macrophyte maximum colonisation depth in Loch Leven, Scotland. Aquatic Botany 91: 75–81.
Straile, D., D. M. Livingstone, G. A. Weyhenmeyer & D. G. George, 2003. The response of freshwater ecosystems to climate variability associated with the North Atlantic oscillation. Geophysical Monographs 134: 263–279.
Townend, J., 2004. Practical statistics for environmental and biological scientists. Wiley, West Sussex.
van Luijn, F., D. T. van der Molen, W. J. Luttmer & P. C. M. Boers, 1995. Influence of benthic diatoms on the nutrient release from sediments of shallow lakes recovering from eutrophication. Water Science and Technology 32: 89–97.
van Nes, E., W. Rip & M. Scheffer, 2007. A theory for cyclic shifts between alternative states in shallow lakes. Ecosystems 10: 17–27.
Wetzel, R. G. & G. E. Likens, 2000. Limnological analyses, 3rd ed. Springer-Verlag, New York.

Hydrobiologia (2012) 681:35–47
DOI 10.1007/s10750-011-0923-x

LOCH LEVEN RESEARCH

# Water quality of Loch Leven: responses to enrichment, restoration and climate change

**L. Carvalho · C. Miller · B. M. Spears · I. D. M. Gunn · H. Bennion · A. Kirika · L. May**

Published online: 3 November 2011


**Abstract** It is usually assumed that climate change will have negative impacts on water quality and hinder restoration efforts. The long-term monitoring at Loch Leven shows, however, that seasonal changes in temperature and rainfall may have positive and negative impacts on water quality. In response to reductions in external nutrient loading, there have been significant reductions in in-lake phosphorus concentrations. Annual measures of chlorophyll *a* have, however, shown little response to these reductions. Warmer spring temperatures appear to be having a positive effect on *Daphnia* densities and this may be the cause of reduced chlorophyll *a* concentrations in spring and an associated improvement in water clarity in May and June. The clearest climate impact was the negative relationship between summer rainfall and chlorophyll *a* concentrations. This is highlighted in extreme weather years, with the three wettest summers having very low chlorophyll *a* concentrations and the driest summers having high concentrations. To predict water quality impacts of future climate change, there is a need for more seasonal predictions from climate models and a greater recognition that water quality is the outcome of seasonal responses in different functional groups of phytoplankton and zooplankton to a range of environmental drivers.

**Keywords** Lake · Recovery · Chlorophyll *a* · Phosphorus · Nitrogen · Temperature · Rainfall · *Daphnia*

Guest editors: L. May & B. M. Spears / Loch Leven: 40 years of scientific research

L. Carvalho (✉) · B. M. Spears · I. D. M. Gunn · A. Kirika · L. May
Centre for Ecology & Hydrology, Penicuik, Midlothian EH26 0QB, UK
e-mail: laca@ceh.ac.uk

C. Miller
School of Mathematics and Statistics, University of Glasgow, University Gardens, Glasgow G12 8QW, UK

H. Bennion
Environmental Change Research Centre, Department of Geography, University College London, Gower Street, London WC1E 6BT, UK

## Introduction

Monitoring at Loch Leven has spanned periods of increasing and decreasing nutrient loads (Bennion et al., 2001; May et al., 2011) and of changing climate (Ferguson et al., 2008). This article examines how the loch has responded to changes in these pressures, and to restoration activities, providing a valuable case-study of how a shallow lake ecosystem reacts to change.

Nutrient enrichment is recognised as one of the most widespread pressures threatening the quality of freshwaters (Carvalho & Moss, 1995; Smith et al., 2006). Over recent decades, great efforts have been made to restore enriched systems, largely through reductions in point-sources of nutrients entering lakes. The recovery of lake ecosystems is not, however, as immediate or as effective as often hoped (Carvalho et al., 1995; Jeppesen et al., 2005; Moss et al., 2005). Attempts to restore lakes are often hindered by internal loading from nutrients that have stockpiled in the lake's sediments (Sas, 1989; Søndergaard et al., 2007; Spears et al., 2007a). More recently, a possible influence of climate change on lake recovery has also been highlighted (Battarbee et al., 2005; Whitehead et al., 2009). The effects of this on lake ecosystems are poorly understood, but it is often assumed that rising temperatures will lead to a deterioration in water quality as this will stimulate phytoplankton growth, particularly that of bloom-forming cyanobacteria (Paerl & Huisman, 2008). However, changes in temperature also affect the growth and reproduction of zooplankton grazers (Elmore, 1983; Hanazato & Yasuno, 1985) and changes in other climate parameters, such as rainfall, may have confounding influences. Long term monitoring records, such as those from Loch Leven, are invaluable in providing observational evidence of how lake ecosystems respond to multiple pressures. They can also help us understand the relative strengths of the key drivers of water quality.

This article examines annual and seasonal trends in the main water quality variables at Loch Leven (phosphorus, nitrogen and chlorophyll *a* concentrations and Secchi depth transparency) in relation to changes in potential key drivers, specifically the zooplankton grazer community, the timing of reductions in external nutrient load and the weather. It examines these changes in relation to water quality targets set by the Loch Leven Area Management Advisory Group (LLAMAG, 1993) and in relation to more recent targets set under the Water Framework Directive (WFD) (European Parliament, 2000).

## Methods

### Site details

Loch Leven is the largest shallow lake in Great Britain (lake area 13.3 $km^2$, mean depth 3.9 m, maximum depth 25.5 m), and is situated in lowland Scotland, UK (56°12′N, 3°22′W; altitude 107 m). Historically, the loch received phosphorus (P)-rich effluent from a woollen mill (Holden & Caines, 1974). Inputs were progressively reduced from a peak in the 1960s and early 1970s and the mill ceased using P-based materials in 1989 (LLCMP, 1999; May et al., 2011). Sewage treatment works within the catchment were also a major source of P to the loch (Bailey-Watts & Kirika, 1987) and, in response to severe eutrophication problems, tertiary treatment and effluent diversion measures were introduced at these works in the 1990s to reduce the nutrient load from this source (May et al., 2011).

As a result of these changes, the total phosphorus (TP) loading, which had risen to 20 t P $y^{-1}$ by 1985, was reduced to 8 t P $y^{-1}$ by 1995 (Bailey-Watts & Kirika 1999; May et al., 2011) and remained at this level in 2005 (May et al., 2011). Since 1995, diffuse nutrient loads from agriculture have also been targeted through the introduction of buffer strips along the Pow Burn although the impact of this is currently unclear (May et al., 2011).

Water quality targets for restoring the ecological health of Loch Leven were originally set by the Loch Leven Area Management Advisory Group (LLAMAG, 1993). The targets were based on the water quality required to support aquatic plants growing to 4.5 m depth, based on historical plant survey records (West, 1910) and using the method outlined by D'Arcy et al. (2006). Those targets are 40 μg $l^{-1}$ annual mean TP, 15 μg $l^{-1}$ annual mean chlorophyll *a* and 2.5 m annual mean Secchi disc depth. The target for TP is consistent with that inferred from a diatom-phosphorus transfer function for the pre-enrichment period of the loch (Bennion et al., 2004).

More recent water quality targets for Loch Leven are being set under the EU Water Framework Directive (European Parliament, 2000), and new UK Environmental Standards for TP for supporting good ecological status under the WFD are also being implemented (UK TAG, 2008). The new TP standards for the loch are more stringent than those set by LLAMAG (1993). The good/moderate (G/M) boundary TP target for a shallow, high alkalinity loch in Scotland, such as Loch Leven, is 32 μg $l^{-1}$ (annual geometric mean) and the moderate/poor (M/P) boundary is 46 μg $l^{-1}$. New European standards for chlorophyll *a* concentrations in lakes have also been developed (Carvalho et al., 2008) and are formally being agreed by the European Commission as part of the Intercalibration process for the WFD. The G/M class boundary for chlorophyll *a* for a shallow, high alkalinity lake, such as Loch Leven, is 7.5 μg $l^{-1}$ (annual arithmetic mean; Carvalho et al., 2006). Loch Leven is, however, near the 3 m mean depth boundary between shallow and very shallow lake types; a site-specific target based on a lake's specific mean depth and alkalinity is, therefore, considered more appropriate. Site-specific chlorophyll *a* targets for Loch Leven would be 11 μg $l^{-1}$ for the G/M boundary and 22 μg $l^{-1}$ for the M/P boundary (Carvalho et al., 2009).

### *Water sampling, storage and analysis*

Loch Leven has been monitored on a weekly to fortnightly basis since 1968. Sampling for water chemistry and plankton was carried out by boat from a mid-basin station, just off Reed Bower. Integrated water samples (between the water surface and about 0.25 m above the lake bed) were collected in duplicate at the site using a weighted polythene tube. As a result of fluctuations in water level, sample depths varied from 3.0 to 3.5 m. Water clarity was measured using a Secchi disk at the Reed Bower station and at the South Deeps when Secchi depth was observed to exceed water depth at Reed Bower. When access to the Reed Bower station was restricted, surface water samples were collected from the sluices at the outflow using a bucket.

Open water crustacean zooplankton samples were collected at the Reed Bower and centre loch sites. Samples were preserved in 4% formaldehyde solution. Sampling and counting methods for estimating *Daphnia* densities are detailed in Gunn et al. (2011).

On return to the laboratory, sub-samples of filtered (Whatman® GF/C) water were taken from each of the duplicate samples for analysis of SRP, nitrate-nitrogen ($NO_3$-N) and soluble reactive silicate (SRSi), and similarly from unfiltered water for analysis of TP. Samples for chlorophyll *a* analysis were prepared by filtering 400 ml of lake water through a GF/C filter. The filter was stored frozen until analysis.

Filtered water was analysed for SRP following the method of Murphy & Riley (1962). TP was determined using a sulphuric acid–potassium persulphate digestion on unfiltered samples to convert all forms of P to SRP. This was then measured using a modified version of Murphy & Riley (1962) as described by Wetzel and Likens (2000). Nitrate was determined using a SEAL AQ2 analyser (SEAL Analytical Limited) by the sulphanilamide/NEDD reaction which produces a reddish-purple dye (HMSO, 1981). SRSi was analysed according to Golterman et al. (1978) using a spectrophotometer fitted with a 10 mm flow-cell.

Chlorophyll *a* was analysed following a methanol extraction method (frozen filters submersed in 90% methanol overnight in a dark fridge) modified from Holm-Hansen & Riemann (1978). The following day, the tubes were centrifuged for 10 min at 2500 r.p.m. Chlorophyll *a* was measured spectrophotometrically at 665 nm with a turbidity correction conducted at 750 nm. The concentration of chlorophyll *a* was determined using a standard calibration equation (APHA, 1992).

Monthly mean air temperature and cumulative rainfall totals were calculated from daily measurements recorded at a meteorological station on the shore. For daily air temperature, an average daily temperature was calculated as [(maximum + minimum)/2]. Data gaps were filled using regression equations relating monthly weather data at Loch Leven with monthly records from the Royal Air Force base at Leuchars, 44 km north-east of Loch Leven (http://www.metoffice.gov.uk/climate/uk/stationdata/).

### *Statistical methods*

Seasonal means were calculated for winter (D, J, F), spring (M, A, M), summer (J, J, A) and autumn (S, O, N) by averaging monthly means for all variables, except rainfall, for which cumulative seasonal rainfall was computed. Annual means were then estimated from seasonal means. This nested approach for estimating monthly, then seasonal, and finally annual means was necessary due to variation in the sampling frequency over the 40-year period, with weekly summer sampling, a common feature in the early monitoring years, biasing averages based on raw sampling data.

For each variable for the trend analysis, two time periods were considered. These were 1968 to 2007 (full 40 years of monitoring) and 1988 to 2007 (last 20 years after major external nutrient load reductions); annual means were also considered for some variables in 2008. In some cases comparisons were also made between the first (1968–1977) and last (1998–2007) decade on record. As a result of the absence of *Daphnia* from the lake between 1968 and June 1970, and different zooplankton sampling sites and methods being used in the early 1970s, trends in *Daphnia* density were investigated from 1975 onwards, only. A natural log transformation was applied to all of the data, except air temperature and cumulative rainfall, to help to stabilize the variance. For *Daphnia,* the constant 0.01 was added before logarithmic transformations were applied, to remove zeros.

Additive and non-parametric regression models (Hastie & Tibshirani, 1990, Bowman & Azzalini, 1997) were developed to model trends and seasonality in the main water quality parameters and potential key drivers of water quality (grazers, air temperature, rainfall). Statistical analyses were carried out to investigate, trends over the 40-year period, seasonal patterns throughout the year and trends over 40 years for each season individually, for each variable of interest. In each model, smooth functions are fitted for each explanatory variable to enable flexibility in the nature of trends detected (i.e. linear, non-linear, non-parametric). For all models, approximate $F$ tests (Hastie & Tibshirani, 1990) were used to test hypotheses concerning the significance of non-parametric trends and to assess whether linear (instead of nonparametric) relationships are appropriate. Analyses were carried out using R software (http://www.r-project.org). Figures and tests for seasonal trends were produced in R using the sm library (Bowman & Azzalini, 1997), in particular, the sm.regression function. Further details of the R code are described in Ferguson et al. (2008). Figures are provided for each season, which display the fitted nonparametric regression line for trend over the 40-year period along with shaded bands to display a reference band that represents 'no effect', i.e. where the curve is expected to lie if there is no evidence of change over time (see Bowman & Azzalini, 1997, for details). For other basic statistics (linear regression, correlation and $t$ tests) analyses were carried out using Minitab version 14.

## Results

### Annual trends in target water quality indicators

Highly significant trends ($P < 0.001$) were observed for all three water quality indicators (TP, chlorophyll $a$, Secchi depth). TP concentrations declined from an annual mean of over 100 μg l$^{-1}$ in the early 1970 s to concentrations below the LLAMAG target of 40 μg l$^{-1}$ in recent years (Fig. 1a). The declining trend was non-linear, with a rapid decline in concentrations in the early 1970s, a slight increase in the early 1990 s and another rapid decline in 2007 and 2008 (to 32 and 33 μg l$^{-1}$, respectively), the latter being at or near the WFD good status target of 32 μg l$^{-1}$.

Chlorophyll $a$ concentrations had a similar rapid decline in concentrations in the early 1970 s, from annual means of over 90 μg l$^{-1}$ to values around

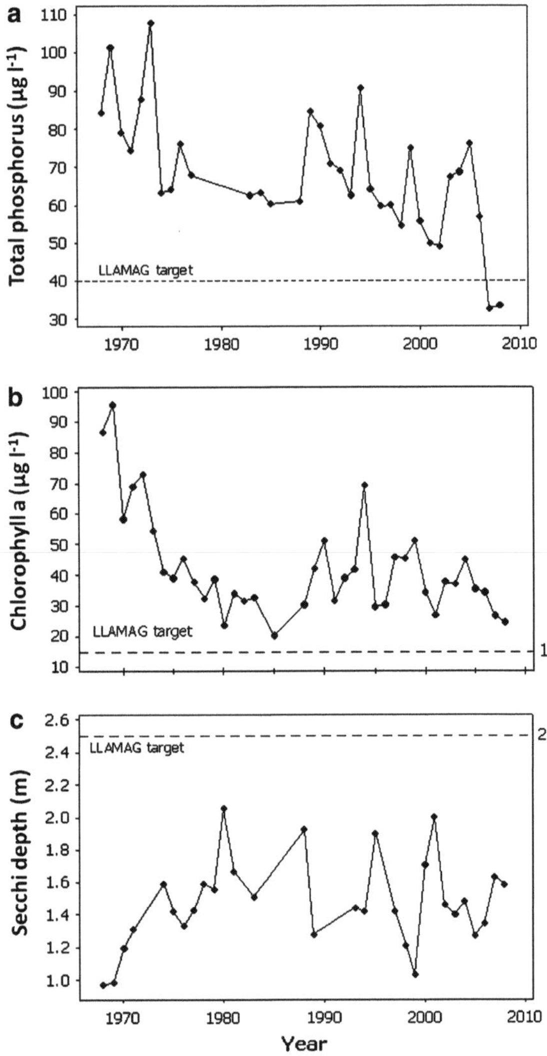


**Fig. 1** Annual mean concentrations of **a** total phosphorus, **b** chlorophyll $a$, and **c** Secchi depth in Loch Leven, 1968–2008, in relation to LLAMAG water quality targets

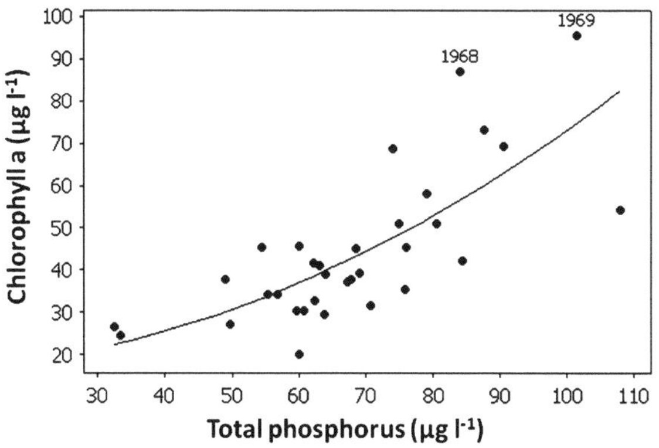


**Fig. 2** Relationship between annual mean chlorophyll *a* and TP concentrations with quadratic regression fit

40 μg $l^{-1}$ by the mid-1970s. Since then, annual mean concentrations have fluctuated, mainly between 30 and 50 μg $l^{-1}$ (Fig. 1b). No significant trend was observed when only the last 20 years (1988–2007) were considered ($P = 0.814$). Chlorophyll *a* levels were relatively low in 2007 (26 μg $l^{-1}$) and 2008 (25 μg $l^{-1}$), but were still well above both LLAMAG and WFD good status targets of 15 and 11 μg $l^{-1}$, respectively. As such, the loch would be classified as having poor ecological status under the WFD. Comparing variability in chlorophyll *a* and TP concentrations (Fig. 2), there is a highly significant positive relationship ($P < 0.001$), indicating that the two are highly inter-dependent. There is, however, some evidence for a levelling off of the relationship at TP concentrations below about 70 μg $l^{-1}$, particularly for the low TP years of 2007 and 2008. The first two years of the monitoring 1968 and 1969 are also large outliers in the regression with more chlorophyll than would be predicted from TP concentrations (Fig. 2).

Secchi disc depths also showed a rapid improvement in the early 1970s, increasing from annual means of around 1.0 m to about 1.5 m by the mid-1970s. Since then, water clarity has been relatively stable with values generally ranging from 1.2 to 1.7 m, although in terms of visibility there have been some particularly poor years (1998, 1999) and particularly good years with annual means of 2.0 m or above (1980, 2000) (Fig. 1c). Secchi depth transparencies in 2007 and 2008 were slightly above average (1.63 and 1.61 m), but were well below the LLAMAG target of 2.5 m.

## Seasonality and seasonal trends in bio-available nutrients

Concentrations of SRP showed significant trends for the spring and winter seasons (Table 1). Trends in spring concentrations have been non-linear, with concentrations declining slightly in the 1980s and increasing slightly from the 1990s, onwards. Winter SRP concentrations show a more consistent linear declining trend. Comparing the seasonality in SRP concentrations for the first (1968–1977) and last (1998–2007) decade of the monitoring period (Fig. 3a), it can be seen that SRP concentrations have declined for most months of the year, with only the spring months showing little change.

Non-parametric regression indicated no significant trends in nitrate-N(N) concentrations for any of the four seasons (Table 1). Winter N concentrations in 1992 (December 1991 to February 1992) were, however, extremely low. Removing this outlier, winter N concentrations showed a highly significant increasing linear trend ($P = 0.003$). Comparing the seasonality in N concentrations for the first (1968–1977) and last (1998–2007) decade of the monitoring (Fig. 3b), concentrations were much higher for the first 6 months

**Table 1** Nonparametric regression test results for seasonal trends in concentrations of Ln soluble reactive phosphorus (SRP), Ln nitrate-N, Ln soluble reactive silicon (SRSi), air temperature, cumulative rainfall, Ln *Daphnia* densities, Ln total phosphorus (TP), Ln chlorophyll*a* and Ln Secchi depth (*p* value for approximate *F* test)

| Variable | Ln SRP | Ln nitrate-N | Ln SRSi | Air Temp. | Rainfall | Ln *Daphnia* | Ln TP | Ln chlorophyll | Ln Secchi |
|---|---|---|---|---|---|---|---|---|---|
| Spring | **0.005** | 0.139 | **0.004** | **0.004** | 0.307 | 0.394 | **0.003** | **0.001** | **0.007** |
| Summer | 0.196 | 0.078 | **0.004** | 0.452 | 0.482 | 0.166 | 0.052 | **0.004** | 0.119 |
| Autumn | 0.127 | 0.577 | 0.701 | **0.005** | 0.248 | 0.671 | 0.472 | 0.217 | 0.067 |
| Winter | **0.029** | 0.359 | **0.042** | **0.029** | **0.040** | 0.253 | 0.306 | **0.023** | 0.195 |

Values in bold indicate a significant non-parametric trend from 1968–2007, except Ln *Daphnia* densities (1975–2007)

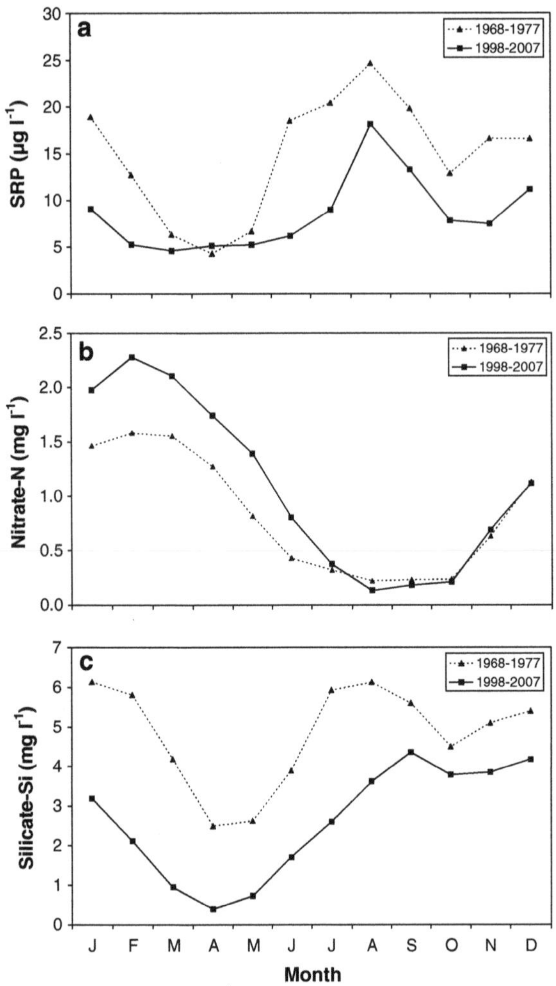


**Fig. 3** Seasonality in monthly mean concentrations of **a** soluble reactive phosphorus (SRP), **b** nitrate-nitrogen, and **c** soluble reactive silicon (SRSi) in Loch Leven for two decades: 1968–1977 and 1998–2007

of the year in the last decade than in the earlier decade. In contrast, there has been little change in N concentrations in the second half of the year, although concentrations in August, September and October were lower in the most recent decade than the first, with levels below analytical detection limits (0.01 mg $l^{-1}$) being recorded more frequently in August.

Soluble reactive silicate concentrations showed highly significant, decreasing trends for spring, summer and winter periods (Table 1). Comparing the seasonality in SRSi concentrations for the first (1968–1977) and last (1998–2007) decade of monitoring (Fig. 3c), it can be seen that concentrations have declined for all months of the year. Since 2000, concentrations in March and April frequently remained below 0.5 mg $l^{-1}$.

## Trends in weather and Daphnia grazers

Air temperature showed a highly significant, increasing linear trend in spring and significant, increasing trends in autumn and winter (Table 1; Fig. 4). Summer was the only season not to have a significant warming trend, although the summers from 2003–2006 were all warmer than average, with 2003 being the warmest summer of the 40-year monitoring period (i.e. 1.6° warmer than the 30 year average of 13.3°C calculated from the 1971–2000 summer means). The most anomalous air temperature was recorded in the winter of 1989, when the mean winter temperature was 5.8°C, almost 3° above the 30 year average of 2.9°C (based on 1971–2000 winter means).

Winter rainfall showed a significant increasing trend over the monitoring period (Table 1; Fig. 5) and an approximate F-test highlighted that a nonparametric trend was not required here and that a linear trend was more appropriate. All seasons, however, showed large variability in rainfall. The years 1990, 1995, 2000 and 2007 had particularly wet winters with over 150 mm more cumulative rainfall than the 30 year average of 293 mm (1971–2000 winter mean). The years 1976, 1996 and 2006 had particularly dry winters with at least 120 mm less cumulative rainfall than the 30 year average. The wettest summers over the 40 year monitoring period were 1985, 1988, 2007 and 2008 with more than 300 mm cumulative rainfall (compared with the 1971–2000 30 year average of 196 mm). The driest summer, by far, was 1995 with only 64 mm of rain.

Since 1975, *Daphnia* densities have shown no significant trends for the four seasonal means (Table 1). However, when densities over the last two decades are compared, it is clear that densities have increased markedly in May, although the difference between the two decades is not quite statistically significant ($P = 0.079$). A general decline in densities is also apparent between July and October when these two decades are compared (Fig. 6).

## Seasonal trends in chlorophyll *a* and secchi depth

Analysis of seasonal trends reveals significant non-linear declining trends for spring, summer and winter chlorophyll *a* concentrations (Table 1),

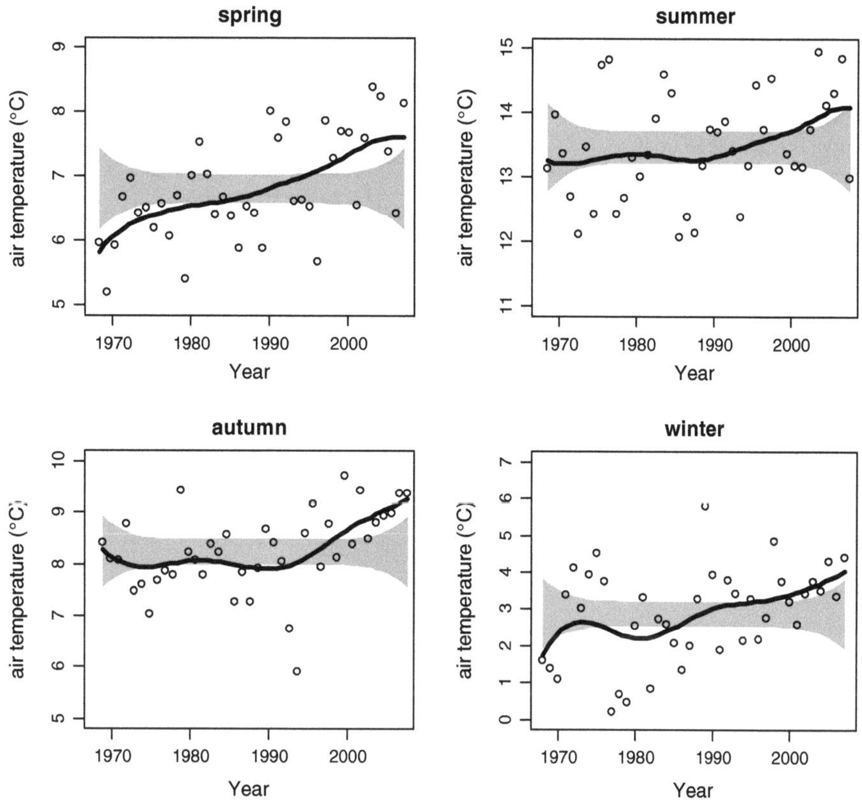


**Fig. 4** Trends in seasonal mean air temperature at Loch Leven, 1968–2007; shaded reference band for 'no effect' shows where the curve is expected to lie if there is no evidence of change over time; note that each plot has a different scale on the *y*-axis

although reductions in the latter two seasons were largely confined to the first decade of monitoring (Fig. 7). If trends are only considered for the last 20 years (1988–2007), only the spring season shows evidence of decreasing concentrations, although this trend is not quite significant ($P = 0.085$). What is even more evident from the raw data are the particularly low chlorophyll *a* concentrations that have been recorded in May since 2000, with concentrations frequently below 10 μg $l^{-1}$. These were significantly different to concentrations of more than 100 μg $l^{-1}$ that were often recorded in this month during the first 6 years (1968–1973) ($P < 0.001$). Low summer chlorophyll *a* concentrations in recent years (1985, 2004, 2007 and 2008) were associated with very wet summers. Similarly some of the driest summers had the highest chlorophyll concentrations (1994, 1995 and 2006). Correlations between log summer rainfall and summer chlorophyll *a* concentrations reveal a changing relationship over the 40 years, with a very weak relationship observed for the first two decades, 1968–1987 ($r = 0.021$, $P = 0.938$) and a significant negative relationship for the last two decades, 1989–2008 ($r = -0.495$, $P = 0.026$). Regression analysis was carried out to consider other potential drivers of summer chlorophyll (nutrients, air temperature and *Daphnia* abundance), but only rainfall showed a significant response.

Analysis of trends in Secchi disc depth by season reveals a highly significant increasing trend in spring, largely driven by increases in the early 1970s (Table 1; Fig. 8). Similar to chlorophyll *a*, the months of May and June have become particularly clear in recent years, with Secchi depths of 3 m or more often being recorded from 2000 onwards, compared with values of less than 1 m for the first three years of the monitoring record.

## Discussion

Annual and seasonal trends in water quality parameters at Loch Leven for the periods 1968–2002 and 1988–2002 were reported by Ferguson et al. (2008). This article brings this analysis up-to-date by considering a full 40 years of monitoring data (1968–2007)

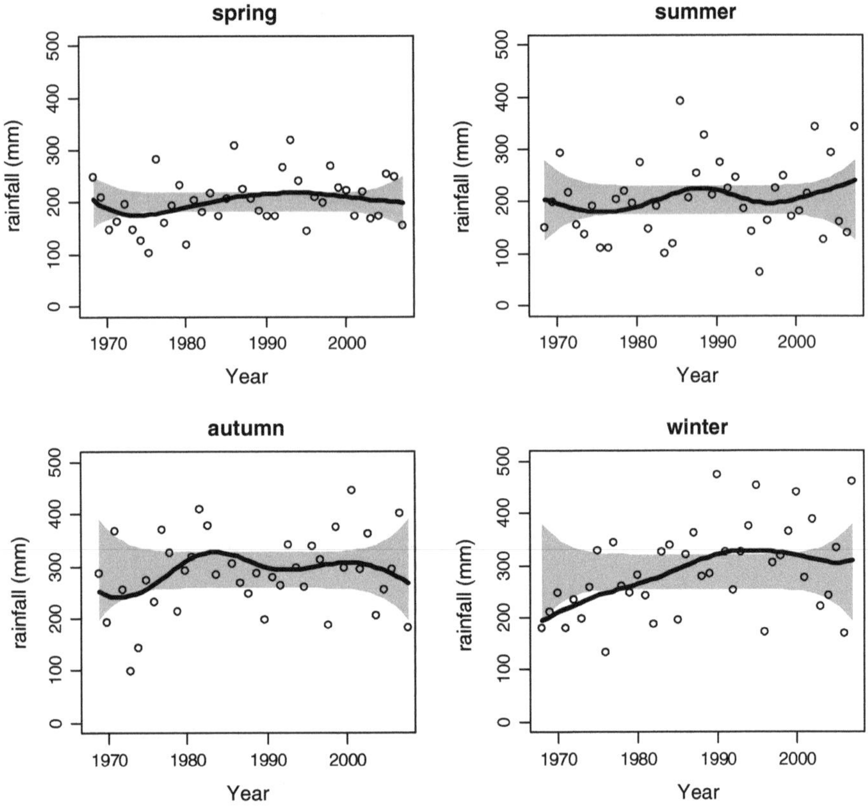


**Fig. 5** Trends in seasonal cumulative rainfall at Loch Leven, 1968–2007; shaded reference band for 'no effect' shows where the curve is expected to lie if there is no evidence of change over time

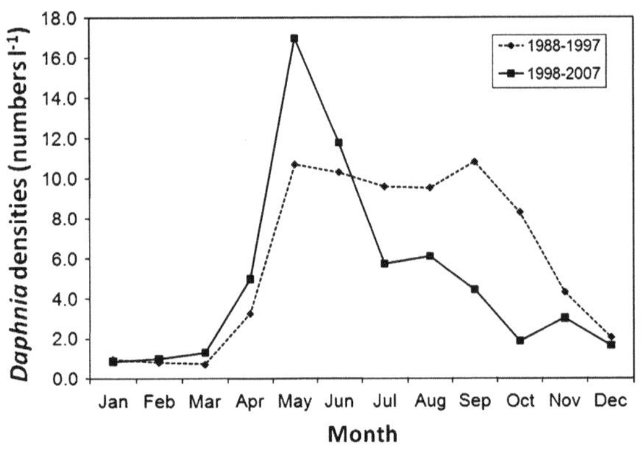


**Fig. 6** Seasonality in *Daphnia* densities in Loch Leven for two decades: 1988–1997 and 1998–2007

and focusing more on changes since 2000. These trends are also discussed in relation to new targets set in response to the WFD.

Annual trends in target water quality parameters

The very abrupt decline in TP and chlorophyll *a* concentrations and the increase in Secchi depth in the early 1970s occurred prior to any reductions in external nutrient loads and coincided with the return of *Daphnia* populations to the lake in 1970 after an absence of 15–20 years (Gunn et al., 2011; May & Spears, 2011). This highlights how significant *Daphnia* grazing is to the water quality of Loch Leven. The impact on TP concentrations, as well as chlorophyll *a* concentrations, indicates an important contribution of phytoplankton cells to the overall TP budget in the water column and hence the potential importance of zooplankton grazing pressure in the maintenance of water quality. Since the mid-1970s, the continuing decline in TP concentrations is most likely a response to the reductions in external loadings of P from the catchment (Bailey-Watts & Kirika, 1999; May et al., 2011). The fact that the annual mean TP concentrations in 2007 and 2008 were below LLAMAG target concentrations of 40 μg $l^{-1}$, and at or near the UK TAG (2008) good/moderate status class boundary, is some evidence that the restoration measures put in place at this site have been a success. However, it is possible that the extremely wet summers of 2007 and 2008 may have contributed to this achievement.

**Fig. 7** Trends in seasonal mean Ln chlorophyll *a* concentrations in Loch Leven, 1968–2007; a shaded reference band for 'no effect' shows where the curve is expected to lie if there is no evidence of change over time

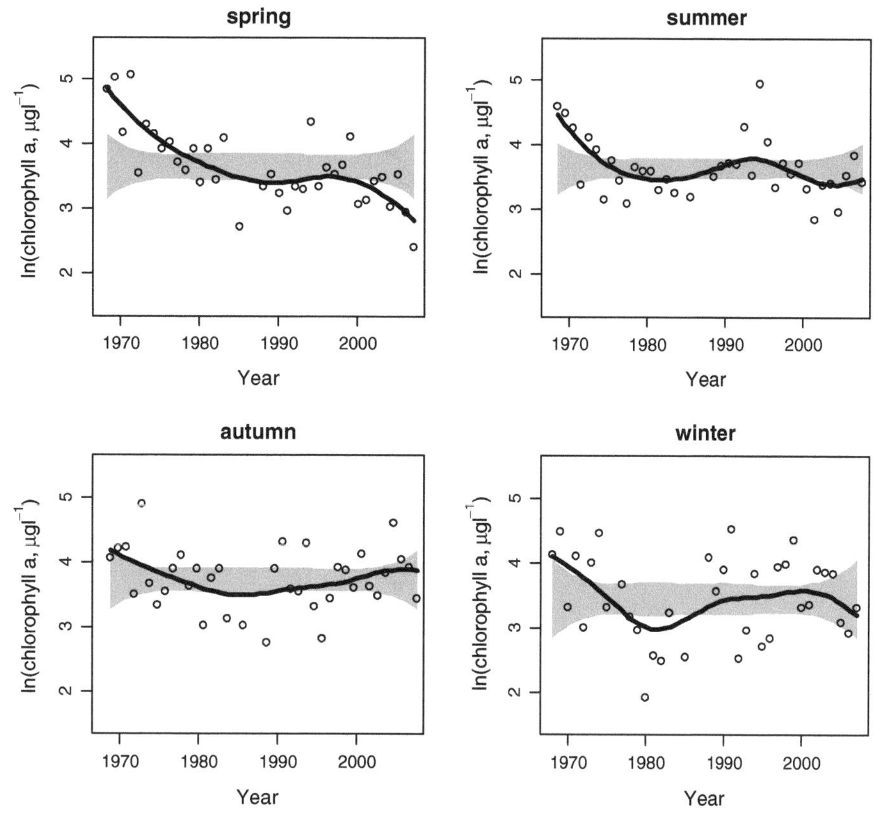


Climatic factors regulating the magnitude of internal loading may have also contributed to the declining trend in annual average TP concentrations (Spears et al., 2011).

In contrast, annual mean chlorophyll *a* concentrations and Secchi depths appear to show a levelling off of a response to the reductions in external P loads that took place since the late 1980s. One possible explanation is that, for most of the monitoring period, P has been in excess for much of the year and has not, therefore, generally limited annual phytoplankton standing crops. A similar lack of response has been observed in other lakes recovering from eutrophication (Jeppesen et al., 2005; Moss et al., 2005) and are commonly due to P release from the sediments sustaining phytoplankton populations. A second possible explanation could be that other factors, such as climate change or grazer densities, may affect the strength of the relationship between annual mean chlorophyll *a* and TP concentrations. This is supported by the TP-chlorophyll regression, which indicates the two biggest outliers in the relationship are associated with 1968 and 1969 when *Daphnia* grazers were absent. In addition to this, it may be that annual measures of chlorophyll *a* mask improvements and/or deteriorations in particular seasons; seasonal trends may be more enlightening, as seasonal processes and consequent relationships between variables (e.g. chlorophyll and P, chlorophyll and grazers) change throughout the year (Ferguson et al., 2009).

Potential for nutrient limitation of phytoplankton

The changing seasonality in SRP concentrations over the 40-year monitoring period highlights a general reduction in concentrations for much of the year, although significant declining trends were only shown for winter. The seasonality of SRP suggests that concentrations are now getting low enough to potentially limit phytoplankton population growth from February through to June. The late summer and autumn peaks in SRP are almost certainly due to internal loading from the P-rich loch sediments (Spears et al., 2007a). The declines observed in the August SRP maximum since the mid-1990s are most likely due to enhanced regulation of P fluxes from the

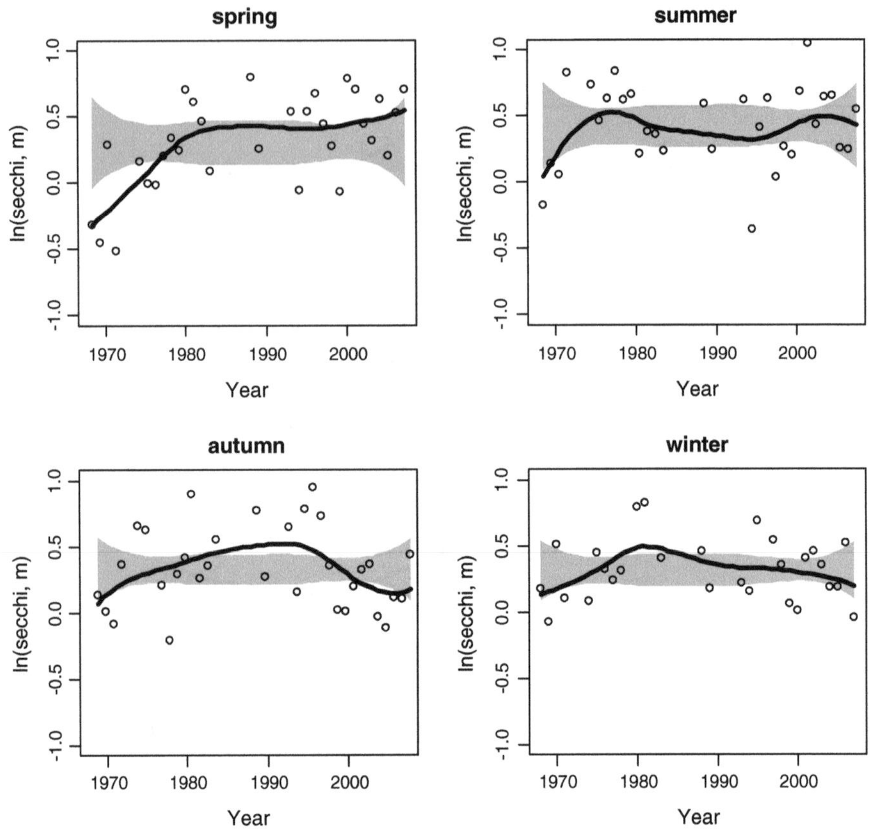


**Fig. 8** Trends in seasonal mean Ln Secchi depth in Loch Leven, 1968–2007; a shaded reference band for 'no effect' shows where the curve is expected to lie if there is no evidence of change over time

sediments, as analyses of the P content of the sediments suggest that the reservoir of sediment-P remains plentiful (Spears et al., 2007a). It is also interesting to note that the late summer/autumn peak has been almost restricted to August over the last decade (1998–2007) compared with June to September in the first decade (1968–1977) of monitoring. This is very comparable to the pattern of recovery observed in another shallow lake, Barton Broad, 10–15 years after sewage diversion (Phillips et al., 2005).

As there is no evidence of increased nitrate-N concentrations in the main inflows (Defew, unpublished data), it is not clear what is driving the increasing winter nitrate-N concentrations in the loch. However, the summer minimum of nitrate-N can probably be explained by two major processes: direct uptake by phytoplankton and denitrification. The latter process, through microbial oxidation of organic carbon, is greatly increased during warmer temperatures (Johnston et al., 1974) and may be one reason why nitrate-N concentrations now reach potentially limiting concentrations for phytoplankton population growth ($<0.1$ mg $l^{-1}$) during late summer. Although this could give nitrogen-fixing cyanobacteria a competitive advantage over other algae (Schindler, 1977), there is no strong evidence of this occurring in Loch Leven. In general, N-fixing cyanobacteria, such as *Anabaena*, tend to peak earlier in the summer in Loch Leven, before N-limitation is most likely to occur.

As we have no SRSi loading data, it is not clear what is causing the significant declines in SRSi concentrations in the loch in spring, summer and winter. SRSi loading would be expected to increase with the increasing winter rainfall, although in-lake concentrations would not be expected to be greatly affected. This suggests that the reductions in lake concentrations may be due more to greater uptake by diatoms. Diatoms do appear to have become the dominant algal group in the loch for much of the year (Carvalho, unpublished data). Only in spring are SRSi concentrations reduced to potentially limiting concentrations for diatom growth (Reynolds, 2006). This may be an important factor in the reduction in spring chlorophyll *a* concentrations, as there has been little change in the other potentially limiting nutrients during this season and diatoms consistently dominate the spring flora.

### Seasonal trends in climate, grazers, chlorophyll *a* and Secchi depth

The seasonal climate trends are similar to those described by Ferguson et al. (2008), i.e. warmer springs and wetter winters. The additional 5 years of monitoring data also reveal that autumn and winter periods are becoming significantly warmer. The results of trend analysis of *Daphnia* densities contrast with the earlier study, with no significant trends apparent since 1975, compared with the highly significant increasing trends recorded from 1968–2002 (Ferguson et al., 2008). This highlights the fact that the latter was driven largely by the absence of *Daphnia* in the first few years of the monitoring period. The only evidence of increasing *Daphnia* densities is for spring, particularly the month of May, with lower densities more common now over much of the summer and autumn (Fig. 6). Gunn et al. (2011) have shown a significant declining trend in mean summer *Daphnia* monthly maxima over the study period. There is, therefore, no evidence that summer *Daphnia* densities have responded positively to the observed increased coverage by macrophytes (Dudley et al., 2011). The lower mean and maxima densities in summer cannot be explained by lower food quantity, as summer chlorophyll trends show no significant decline since 1988. The decline over the last decade may, therefore, be due to either poorer quality food for *Daphnia* (e.g. inedible or toxic algae) or to increased predation by zooplanktivorous fish.

The enhanced *Daphnia* populations in May are certainly one of the most plausible reasons for the reductions in May chlorophyll *a* concentrations and improvements in May/June water clarity, although declining SRSi concentrations limiting diatom growth are likely to be a major contributory factor too.

Variability in summer chlorophyll *a* concentrations over the last two decades does appear to be increasingly in response to variability in climate, specifically summer rainfall. The best summers, in terms of reduced chlorophyll *a* concentrations, are associated with particularly wet summers (1985, 2007 and 2008) and the worst summers with particularly dry conditions (1994, 1995 and 2006).

### Climate change and lake recovery

Detailed analyses of the seasonal trends indicate responses to changes in both nutrients and climate. There appear to have been significant responses in in-lake P concentrations to the reductions in point-source nutrient loading. In terms of chlorophyll *a* concentrations, warmer temperatures potentially have direct physiological effects (e.g. enhanced growth rates) on plankton communities (Elmore, 1983; Hanazato & Yasuno, 1985; Reynolds, 2006), and it is generally assumed that this will result in an increase in phytoplankton or zooplankton abundance. The negative relationship observed at Loch Leven, with reduced spring chlorophyll *a* and warmer spring water temperatures, suggests an indirect response, probably due to a strong positive effect of water temperature on grazer densities (Ferguson et al., 2007). The consequent increases in May/June water clarity may be a key factor driving the recolonisation of deeper water by submerged aquatic plants (Dudley et al., 2011; May & Carvalho, 2010) and consequent benefits to associated biodiversity (Carss et al., 2011; Gunn et al., 2011; Winfield et al., 2011), although other factors, such as changes in wind disturbance of sediments, may also be important (Spears & Jones, 2010). The most obvious climate impact on water quality was the beneficial effect of very wet summers, which in recent decades has been significantly associated with low summer chlorophyll *a* concentrations. The most direct cause of this is probably increased phytoplankton loss processes through enhanced flushing from the loch (Bailey-Watts et al., 1990). This is likely to particularly affect slow-growing phytoplankton, such as bloom-forming cyanobacteria (Reynolds & Lund, 1988). Increased flushing will also help the long-term success of restoration, by exporting P released from internal sediment sources during the summer via the outflow (Spears et al., 2007b). As the outflow of Loch Leven is controlled, increased summer rainfall may also be partly retained in the loch, which may simply dilute loch phytoplankton populations.

The long-term research at Loch Leven is helping us understand how shallow lakes respond to both lake restoration and climate change. It is usually assumed that climate change will have negative impacts on water quality, but this research suggests that some seasonal changes in air temperature and rainfall patterns may actually have positive effects. To predict water quality responses to future climate change, there is a clear need for better seasonal predictions from climate models and a greater understanding of how different functional groups of phytoplankton and zooplankton respond to these seasonal changes.

**Acknowledgements** Most of the sample collection and analysis for this research was funded by the Natural Environment Research Council with additional support from the European Union (FP6 Integrated Project 'Euro-limpacs: European project to evaluate impacts of global change on freshwater ecosystems' GOCE-CT-2003-505540). The authors also gratefully acknowledge Kinross Estates for providing access to the loch and for assistance with fieldwork. Loch Leven is part of the UK Environmental Change Network (http://www.ecn.ac.uk/).

## References

APHA, 1992. Standard Methods for the Examination of Water and Wastewater, 18th ed. American Public Health Association, Washington.

Bailey-Watts, A. E. & A. Kirika, 1987. A re-assessment of the phosphorus inputs to Loch Leven (Kinross, Scotland): rationale and an overview of results on instantaneous loadings with special reference to runoff. Earth Sciences 78: 351–367.

Bailey-Watts, A. E. & A. Kirika, 1999. Poor water quality in Loch Leven (Scotland) in 1995, in spite of reduced phosphorus loadings since 1985: the influences of catchment management and inter-annual weather variation. Hydrobiologia 403: 135–151.

Bailey-Watts, A. E., A. Kirika, L. May & D. H. Jones, 1990. Changes in phytoplankton over various time scales in a shallow, eutrophic: the Loch Leven experience with special reference to the influence of flushing rate. Freshwater Biology 23: 85–111.

Battarbee, R. W., N. J. Anderson, E. Jeppesen & P. R. Leavitt, 2005. Combining palaeolimnological and limnological approaches in assessing lake ecosystem response to nutrient reduction. Freshwater Biology 50: 1772–1780.

Bennion, H., J. Fluin, P. Appleby & B. Ferrier, 2001. Palaeolimnological investigation of Scottish freshwater lochs. Final Report to SNIFFER No. SR(00)02. ENSIS Ltd, University College London, London.

Bennion, H., J. Fluin & G. L. Simpson, 2004. Assessing eutrophication and reference conditions for Scottish freshwater lochs using subfossil diatoms. Journal of Applied Ecology 41: 124–138.

Bowman, A. & A. Azzalini, 1997. Applied Smoothing Techniques for Data Analysis. Clarendon Press, Oxford.

Carss, D., B. M. Spears, L. Quinn & R. Cooper, 2011. Long-term variations in waterfowl populations in Loch Leven: identifying discontinuities between local and national trends. Hydrobiologia. doi:10.1007/s10750-011-0927-6.

Carvalho, L. & B. Moss, 1995. The current status of a sample of English Sites of Special Scientific Interest subject to eutrophication. Aquatic Conservation 5: 191–204.

Carvalho, L., M. Beklioglu & B. Moss, 1995. Changes in a deep lake following sewage diversion—a challenge to the orthodoxy of external phosphorus control as a restoration strategy? Freshwater Biology 34: 399–410.

Carvalho, L., G. Phillips, S. Maberly & R. Clarke, 2006. Chlorophyll and phosphorus classifications for UK Lakes. Final Report to SNIFFER (Project WFD38), Edinburgh, October 2006: 81 pp.

Carvalho, L., A. Solimini, G. Phillips, M. van den Berg, O.-P. Pietiläinen, A. Lyche, S. Poikane & U. Mischke, 2008. Chlorophyll reference conditions for European Intercalibration lake types. Aquatic Ecology 42: 203–211.

Carvalho, L., A. G. Solimini, G. Phillips, O.-P. Pietiläinen, J. Moe, A. C. Cardoso, A. Lyche Solheim, I. Ott, M. Søndergaard, G. Tartari & S. Rekolainen, 2009. Site-specific chlorophyll reference conditions for lakes in Northern and Western Europe. Hydrobiologia 633: 59–66.

D'Arcy, B. J., L. May, J. Long, I. R. Fozzard, S. Greig & A. Brachet, 2006. The restoration of Loch Leven, Scotland, UK. Water Science Technology 53: 183–191.

Dudley, B., I. D. M. Gunn, L. Carvalho, I. Proctor, M. T. O'Hare, K. J. Murphy & A. Milligan, 2011. Changes in aquatic macrophyte communities in Loch Leven: evidence of recovery from eutrophication? Hydrobiologia doi: 10.1007/s10750-011-0924-9.

Elmore, J. L., 1983. The influence of temperature on egg development times of three species of *Diaptomus* from subtropical Florida. American Midland Naturalist 109: 300–308.

European Parliament, 2000. Directive 2000/60/EC of the European Parliament and of the Council of 23 October 2000, establishing a framework for community action in the field of water policy. Official Journal of the European Communities L327/1: 1–72.

Ferguson, C. A., E. M. Scott, A. W. Bowman & L. Carvalho, 2007. Model comparison for a complex ecological system. Journal of the Royal Statistical Society Series A 170: 691–711.

Ferguson, C. A., L. Carvalho, E. M. Scott, A. W. Bowman & A. Kirika, 2008. Assessing ecological responses to environmental change using statistical models. Journal of Applied Ecology 45: 193–203.

Ferguson, C. A., A. W. Bowman, E. M. Scott & L. Carvalho, 2009. Multivariate varying-coefficient models for an ecological system. Environmetrics 20: 460–476.

Golterman, H. L., R. S. Clymo & M. A. M. Ohnstad, 1978. Methods for Physical and Chemical Analysis of Freshwaters, 2nd ed. Blackwell Scientific, Oxford.

Gunn, I. D. M., M. T. O'Hare, P. S. Maitland & L. May, 2011. Long-term trends in Loch Leven invertebrate communities. Hydrobiologia. doi:10.1007/s10750-011-0926-7.

Hanazato, T. & M. Yasuno, 1985. Effect of temperature in the laboratory studies on growth, egg development and first parturition of five species of Cladocera. Japanese Journal of Limnology 46: 185–191.

Hastie, T. & R. Tibshirani, 1990. Generalized Additive Models. Chapman and Hall, London.

HMSO, 1981. Methods for the Examination of Waters & Associated Materials: Book Number 40. Oxidised Nitrogen in Waters. HMSO, London.

Holden, A. V. & L. A. Caines, 1974. Nutrient chemistry of Loch Leven, Kinross. Proceedings of the Royal Society of Edinburgh, B 74: 101–121.

Holm-Hansen, O. & B. Riemann, 1978. Chlorophyll *a* determination: improvements in methodology. Oikos 30: 438–447.

Jeppesen, E., M. Søndergaard, J. P. Jensen, K. Havens, O. Anneville, L. Carvalho, M. F. Coveney, R. Deneke, M. T. Dokulil, B. Foy, D. Gerdeaux, S. E. Hampton, S. Hilt, K. Kangur, J. Kohler, E. H. H. R. Lammens, T. L. Lauridsen, M. Manca, M. R. Miracle, B. Moss, P. Noges, G. Persson, G. Phillips, R. Portielje, S. Romo, C. L. Schelske, D. Straile, I. Tatrai, E. Willen & M. Winder, 2005. Lake responses to reduced nutrient loading—an analysis of contemporary long-term data from 35 case studies. Freshwater Biology 50: 1747–1771.

Johnston, D. W., A. J. Holding & J. E. McCluskie, 1974. Preliminary comparative studies of denitrification and methane production in Loch Leven, Kinross, and other freshwater lakes. Proceedings of the Royal Society of Edinburgh, B 74: 123–134.

LLAMAG, 1993. Loch Leven Area Management Advisory Group Report. Scottish Natural Heritage, Battleby, Perth, UK.

LLCMP, 1999. Loch Leven Catchment Management Plan: The Report of the Loch Leven Area Management Advisory Group.

May, L. & L. Carvalho, 2010. Maximum growing depth of macrophytes in Loch Leven, Scotland, United Kingdom, in relation to historical changes in estimated phosphorus loading. Hydrobiologia 646: 123–131.

May, L. & B. M. Spears, 2011. A history of scientific research at Loch Leven, Kinross, Scotland. Hydrobiologia. doi: 10.1007/s10750-011-0929-4.

May, L. L. H. Defew, H. Bennion & A. Kirika, 2011. Historical changes (1905–2005) in external phosphorus loads to Loch Leven, Scotland, UK. Hydrobiologia. doi:10.1007/s10750-011-0922-y.

Moss, B., T. Barker, D. Stephen, A. Williams, D. Balayla, M. Beklioglu & L. Carvalho, 2005. Consequences of reduced nutrient loading on a lake system in a lowland catchment: deviations from the norm? Freshwater Biology 50: 1687–1705.

Murphy, J. & J. P. Riley, 1962. A modified single solution method for the determination of phosphate in natural waters. Analytica Chemica Acta 27: 31–36.

Paerl, H. W. & J. Huisman, 2008. Blooms like it hot. Science 320: 57–58.

Phillips, G., A. Kelly, J. Pitt, R. Sanderson & E. Taylor, 2005. The recovery of a very shallow eutrophic lake, 20 years after the control of effluent derived phosphorus. Freshwater Biology 50: 1628–1638.

Reynolds, C. S., 2006. The Ecology of Phytoplankton. Cambridge University Press, Cambridge.

Reynolds, C. S. & J. W. G. Lund, 1988. The phytoplankton of an enriched, soft-water lake subject to intermittent hydraulic flushing (Grasmere, English Lake District). Freshwater Biology 19: 379–404.

Sas, H., 1989. Lake Restoration by Reduction of Nutrient Loading. Academia Verlag Richartz GmbH, St. Augustin.

Schindler, D. W., 1977. The evolution of phosphorus limitation in lakes. Science 195: 260–262.

Smith, V. H., S. B. Joye & R. W. Howarth, 2006. Eutrophication of freshwater and marine ecosystems. Limnology and Oceanography 51: 351–355.

Søndergaard, M., E. Jeppesen, T. Lauridsen, S. C. H. van Nes, R. Roijackers, E. Lammens & R. Portielje, 2007. Lake restoration: successes, failures and long-term effects. Journal of Applied Ecology 44: 1095–1105.

Spears, B. M. & I. D. Jones, 2010. The long-term (1979–2005) effects of the North Atlantic Oscillation Index on wind-induced wave mixing in Loch Leven (Scotland). Hydrobiologia 646: 49–59.

Spears, B. M., L. Carvalho, R. Perkins, A. Kirika & D. M. Paterson, 2007a. Sediment P cycling in a large shallow lake: spatio-temporal variation in phosphorus pools and release. Hydrobiologia 584: 37–48.

Spears, B. M., L. Carvalho & D. M. Paterson, 2007b. Phosphorus partitioning in a shallow lake: implications for water quality management. Water and Environment Journal 21: 47–53.

Spears B. M., L. Carvalho, R. Perkins, A. Kirika & D. M. Paterson, 2011. Long-term variation and regulation of internal phosphorus loading in Loch Leven Hydrobiologia. doi: 10.1007/s10750-011-0921-z.

UK TAG, 2008. UK Environmental Standards and Conditions (Phase 2). Final Report (SR1–2007), March 2008. SNIFFER, Edinburgh.

West, G., 1910. A further contribution to a comparative study of the dominant phanerogamic and higher cryptogamic flora of aquatic habitat in Scottish lakes. Proceedings of the Royal Society of Edinburgh 30: 65–181.

Wetzel, R. G. & G. E. Likens, 2000. Limnological Analyses, 3rd ed. Springer, New York.

Whitehead, P. G., R. L. Wilby, R. W. Battarbee, M. Kernan & A. J. Wade, 2009. A review of the potential impacts of climate change on surface water quality. Hydrological Sciences 54: 101–123.

Winfield, I. J., C. E. Adams, J. D. Armstrong, R. Gardiner, A. Kirika, J. Montgomery, B. M. Spears, D. C. Stewart, J. E. Thorpe & W. Wilson, 2011. Changes in the fish community of Loch Leven: untangling anthropogenic pressures. Hydrobiologia. doi:10.1007/s10750-011-0925-8.

Hydrobiologia (2012) 681:49–57
DOI 10.1007/s10750-011-0924-9

LOCH LEVEN RESEARCH

# Changes in aquatic macrophyte communities in Loch Leven: evidence of recovery from eutrophication?

B. Dudley · I. D. M. Gunn · L. Carvalho · I. Proctor · M. T. O'Hare · K. J. Murphy · A. Milligan

Published online: 3 November 2011


**Abstract** This article assesses changes in the macrophyte community of Loch Leven over a period of 100 years. Evidence is presented that shows that these changes are associated with eutrophication and with subsequent recovery from eutrophication when anthropogenic nutrient inputs to the loch were reduced. This study uses macrophyte survey data from 1905, 1966, 1972, 1975, 1986, 1993, 1999 and 2008. In each of the four most modern surveys, the loch was divided into 19 sectors, each with at least one transect ranging from the shallowest to the deepest occurrence of macrophytes. From these data, a range of indicators of recovery were derived at the whole lake scale: the relative abundance of taxa, taxon richness and evenness. All of these metrics showed an improvement since 1972. Species richness, measured at the scales of survey sector and individual samples, also appeared to have increased in recent years. All of these measures, coupled with ordination of the presence/absence composition data from all survey years, indicate that the macrophyte community in the loch is recovering towards the state that was recorded in 1905.

**Keywords** Lake · Plant diversity · Phosphorus · Growing depth · Charophyte · *Potamogeton*

Guest editors: L. May & B. M. Spears / Loch Leven: 40 years of scientific research

B. Dudley (✉) · I. D. M. Gunn · L. Carvalho · I. Proctor · M. T. O'Hare
Centre for Ecology & Hydrology, Bush Estate, Penicuik, Midlothian EH26 0QB, UK
e-mail: bedu@ceh.ac.uk

K. J. Murphy · A. Milligan
Division of Environmental & Evolutionary Biology, University of Glasgow, University Avenue, Glasgow G12 8QQ, UK

## Introduction

Loch Leven has been a focus for ecological and water quality research since the late 1960s. Over this period there have been problems with eutrophication caused by industrial, agricultural and sewage effluents, with nutrient inputs reaching a peak in the 1980s (May & Spears, 2011). The resultant deterioration in water quality was reflected in declines in the abundance, diversity and maximum growing depth of macrophytes over this period (Jupp et al., 1974). Since then, external inputs of phosphorus (P) to the loch have been reduced (May & Spears, 2011) and there has been a significant, associated, reduction in P concentrations in the loch water over the last decade (Ferguson et al., 2008; Carvalho et al., 2011).

Macrophytes are generally considered to be good indicators of water quality, particularly in relation to eutrophication pressures (Schaumburg et al., 2004; Søndergaard et al., 2005; Penning et al., 2008a, b). They play a key role in the physical structuring of shallow lakes, providing important habitat for invertebrates, fish and birds (Jeppesen et al., 1998; Warfe & Barmuta, 2006). The US EPA Lake and Reservoir Bioassessment and Biocriteria protocols comment that: "Macrophytes respond more slowly to environmental changes than do phytoplankton or zooplankton and might be better integrators of overall environmental conditions. This would allow a single sampling event per year, during the time of maximum abundance of macrophytes. Both floating leaved and emergent plants can be assessed from aerial photographs, which permit estimates of total area covered and percent cover (density) within stands" (US Environmental Protection Agency, 1998). It is, therefore, timely to examine changes in the macrophyte community of Loch Leven, over both the eutrophication and recovery period, and put these changes into the context of the historical past (West, 1910).

Although earlier records exist for individual species, aquatic macrophytes were first surveyed systematically at Loch Leven by West in 1905 (West, 1910), as part of the bathymetric survey of Scottish Lochs (Murray & Pullar, 1910). Data from the 1905 survey form an historical baseline that more recent data can be compared to, including those from the surveys of Jupp et al. (1974), Britton (1975), Robson (1986), Murphy & Milligan (1993) and Griffin & Milligan (1999). These data can also be compared to macrophyte data collected in 2008 to determine whether recent changes within the macrophyte community continue to track the loch's recovery from eutrophication. This article considers the following questions: (1) Do long-term changes in the species composition of aquatic plants reflect the well-documented eutrophication and subsequent recovery from eutrophication at this site? (2) Has the aquatic plant community returned to a state that is comparable to that observed in 1905?

## Methods

### Site description

Loch Leven is a shallow lake (mean depth 3.9 m; maximum depth about 25 m), with a surface area of 13.3 km$^2$. It is located near the town of Kinross, in the central lowlands of Scotland, UK. The structure and physical environment of the loch are described in detail by Smith (1974). Further details about the catchment, nutrient inputs and historical alterations can be found in May & Spears (2011) and May et al. (2011).

**Table 1** Sources of macrophyte data showing year of survey, total number of samples collected and literature source

| Year | Samples | Source |
|---|---|---|
| 1905 | Unknown | West (1910) |
| 1966 | 853 | Jupp et al. (1974) |
| 1972 | 744 | Jupp et al. (1974) |
| 1975 | 556 | Britton (1975) |
| 1986 | 190 | Robson (1986) |
| 1993 | 233 | Murphy & Milligan (1993) |
| 1999 | 233 | Griffin & Milligan (1999) |
| 2008 | 255 | CEH internal study |

### Data selection

To assess long-term, community-wide changes in the macrophyte community, taxonomic occurrence data were selected on the basis of comparability and completeness. As a result, many historical records, including some of those described by Jupp et al. (1974), were excluded from the analyses because they did not form part of a comprehensive lake survey. Apart from Jupp et al. (1974), most of the data used in this study are from unpublished reports (Table 1). Palaeobotanical data were also available for Loch Leven (Salgado et al., 2009).

The survey conducted in 1905 (West, 1910) is the earliest comprehensive dataset for macrophytes at Loch Leven. This survey gives information on which plants were found although, in contrast to the other studies, it recorded only presence/absence data and the sampling method was not noted. For this reason, these data were only suitable for assessing change in terms of the presence/absence of species.

### Sampling methods

Taxonomic occurrence data were used only from surveys for which the sampling methods were generally consistent. The main differences between surveys were in the total number of samples taken (Table 1), the length of drag when sampling, and the size and shape of sampling rake used.

All of the surveys (apart from 1905) were conducted along pre-determined transects using a boat.

Since 1986, the lake has been divided into 'sectors' that have been sampled consistently in terms of transect and sampling locations. For all data collected since 1986, it has been possible to assign a sector to each sample. This has enabled analysis of diversity across a range of spatial scales.

Samples of the aquatic vegetation at intervals along these transects were obtained using a double headed rake. The rake was constructed from two garden rake heads joined back-to-back with wire mesh attached to their surfaces. Although the basic design of the sampling rake remained constant in most of the surveys, some aspects of the design changed among surveys. For example, Jupp et al. (1974) explained that, in the 1972 survey, drag-rakes varied between transects in relation to their weight (1,240–1,923 g), width (19–28.7 cm), prong length (14–19 cm), number of prongs per side (8–12), prong shape (straight to strongly curved) and screen mesh size (1–2.5 cm). Such detail is not always given in the reports of subsequent surveys.

Drag distances also varied among surveys. In the earlier modern surveys (1966, 1972 and 1974), the boat was kept in constant motion and the sampling rake was allowed to drag along the bottom for 50–100 m (Jupp et al., 1974). In 1986 sampling distances of between 5 and 129 m were used (Robson, 1990). In all of the surveys since then, drag distances of only 2 m have been used (Murphy & Milligan, 1993; Griffin & Milligan, 1999; this study).

In some of the surveys, a bathyscope was also used to observe underwater plants. In the analyses, the taxa seen using this method were given equal consideration to those sampled with the rake. In many of the surveys, a measure of abundance comprising either a weighting or some semi-quantitative measure was given to either individual taxa or the sample as a whole. However, these measures of abundance have not been used as part of the analyses presented here, because it was judged that these measures were not sufficiently similar across the surveys. At the sample level (either bathyscope or rake), only the presence/absence data have been used for this study.

## Taxa aggregation and exclusion

Occurrence records for some taxa were aggregated. This was either because they had been previously aggregated in some of the surveys, or because it was judged that these taxa may have been confused because of their similarity. These aggregates included all *Chara* species, all *Tolypella* and *Nitella* species, *Potamogeton berchtoldii/P. pusillus, P. filiformis/P. pectinatus*, all *Ranunculus* taxa, and *Myriophyllum alterniflorum/M. spicatum*.

Some taxa were excluded from the analyses because they did not seem to have been recorded in a consistent manner. This group included all lemnids (free-floating plants), bryophytes, algae other than charophytes, the floating-leaved *Nymphaea alba*, and emergent vegetation. The remainder comprised submerged vascular plants, with some aggregations, and charophytes. A total of 18 taxa were included in the analyses (Table 2).

## Indicators of recovery

A number of measures were used to investigate temporal and spatial changes within the macrophyte community. The simplest of these was loch-wide richness, which was calculated as the total number of taxa (as defined above) found in any single survey.

Evenness was calculated using Simpson's index ($E_{1/D}$), which was calculated according to Magurran (2004). This measure takes values of between 0 and 1, where a value of 1 implies equal numbers of individuals of all taxa within the population, and an evenness of 0 implies only one taxon. Evenness was calculated for all survey years except 1905, for which no quantitative data were available.

Species counts at smaller spatial scales were examined by comparing the means, medians and 10th, 25th, 75th and 90th percentiles of the number of taxa found in a single sample for years when the sampling methods were considered to be equivalent, i.e. those where there had been equal sampling effort per sample. These were the years 1993, 1999 and 2008. Similarly, the same statistics were compared for the number of taxa found in a survey sector for the years when it was considered that the likelihood of finding all taxa in a sector was equivalent, i.e. there had been equal sampling effort per sector. These were the years 1986, 1993, 1999 and 2008.

A Euclidean multi-dimensional scaling (MDS) analysis was conducted using the R software package (R Development Core Team, 2008). This indirect gradient analysis produces a two-dimensional ordination of compositional change over the survey period.

**Table 2** Taxa included in the analyses of macrophyte data from Loch Leven, including year of survey and an indication of relative abundance, where available abundance data are standardised point frequency, expressed as a percentage

| Taxa | 1905 | 1966 | 1972 | 1975 | 1986 | 1993 | 1999 | 2008 |
|---|---|---|---|---|---|---|---|---|
| *Chara* spp. | x | 97 | 30 | 37 | 56 | 56 | 52 | 22 |
| *Potamogeton berchtoldii/pusillus* | x | | | 1 | | 1 | 8 | 21 |
| *Nitella/Tolypella* | x | 1 | 13 | 28 | | | | 19 |
| *Callitriche hermaphroditica* | x | | 3 | 1 | 3 | 6 | 10 | 14 |
| *Potamogeton perfoliatus* | x | | 1 | 1 | 6 | 7 | 7 | 7 |
| *Potamogeton filiformis/pectinatus* | x | 1 | 25 | 28 | 21 | 5 | 19 | 7 |
| *Elodea canadensis* | x | 2 | 1 | 2 | 2 | 6 | 1 | 6 |
| *Zannichellia palustris* | | 1 | 30 | 5 | 13 | 21 | 2 | 5 |
| *Eleocharis acicularis* | x | | | 2 | | 1 | 2 | 1 |
| *Myriophyllum* spp. | x | 1 | 1 | | | | | 1 |
| *Potamogeton praelongus* | x | | | | | | | 1 |
| *Ranunculus* spp. | x | | | | | | | 1 |
| *Littorella uniflora* | x | | 1 | 1 | 2 | 1 | 2 | 1 |
| *Potamogeton crispus* | | 2 | 1 | 1 | 1 | 2 | | |
| *Potamogeton gramineus* | x | | | | | | | |
| *Potamogeton lucens* | x | | | | | | | |
| *Potamogeton obtusifolius* | x | | 1 | | | | | |
| *Potamogeton* x *zizii* | x | | | | | | | |

NB ‘x’ indicates that no abundance information is available

The presence/absence data from all surveys were used for this analysis.

Maximum growing depth

The maximum depth to which macrophytes grow in Loch Leven is discussed in an historical context by May & Carvalho (2010). Data from this article were combined with the deepest observation of rooted plants from the 2008 survey to examine patterns of change in maximum growing depth over time.

## Results

Long-term variation in macrophyte community composition

The results of the analyses showed that both the taxa present, and their relative abundances, had varied considerably over time. Of the 18 taxa studied, three *Potamogeton* species were found in 1905 and have not been recorded since (i.e. *P. gramineus*, *P. lucens* and the hybrid of *P. gramineus* x *lucens*, also known as *P.* x *zizii*). Another *Potamogeton* species, *P. obtusifolius*, was recorded in 1905 and 1972, but has not been found since. *Potamogeton praelongus* and *Ranunculus* spp. were observed in 2008, but have not been seen since 1905. Similarly *Potamogeton berchtoldii/pusillus* was recorded in 1905, then seen rarely in 1975 and 1993, before returning in some abundance in 1999 and 2008. Only two of the 18 taxa were not found in 1905. These were *Potamogeton crispus*, which was observed consistently from 1966 to 1993 but not since, and *Zannichellia palustris*, which has been found in every survey since 1966, but now appears to be declining.

Long-term variation in indices of diversity

At the whole lake scale, the values of all of the indicators of diversity examined were greater in 2008 than in any of the previous years, apart from 1905. Taxon richness was highest in 1905 (16 taxa), and lowest in 1966 (7 taxa) (Fig. 1a). Taxon richness was 11 in 1972 and 1975, then 8, 10, 9 and 13 in 1986, 1993, 1999 and 2008, respectively. Evenness was

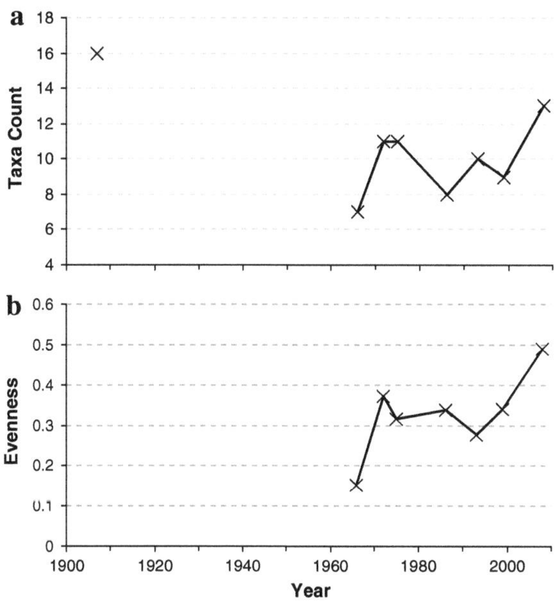


**Fig. 1** Counts of taxa (**a**) and evenness of taxa distribution (**b**) from all Loch Leven macrophyte surveys; see "Methods" section for description of taxa

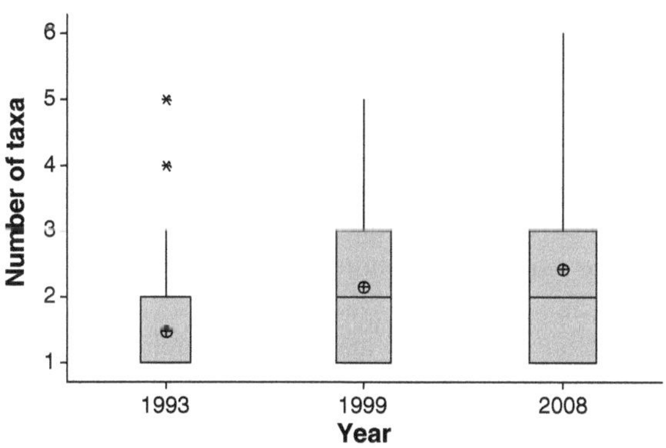


**Fig. 2** Number of taxa found in each sample in 1993, 1999 and 2008; data are summarised as *boxplots*, including the median, 25th, 75th and 90th percentile. Note that, for 1993, the median was 1. Outliers (*stars*) and means (*crossed circles*) are also shown

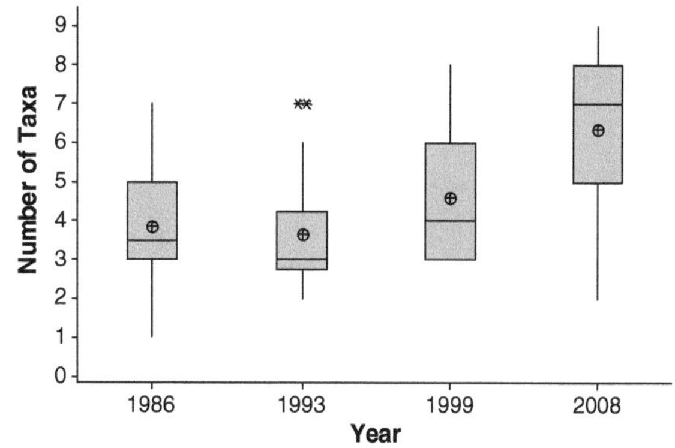


**Fig. 3** Number of macrophyte taxa found in each sector during surveys of Loch Leven in 1986, 1993, 1999 and 2008; data are summarised as box plots showing the median and 10th, 25th, 75th and 90th percentiles. Outliers (*stars*) and means (*crossed circle*) are also shown

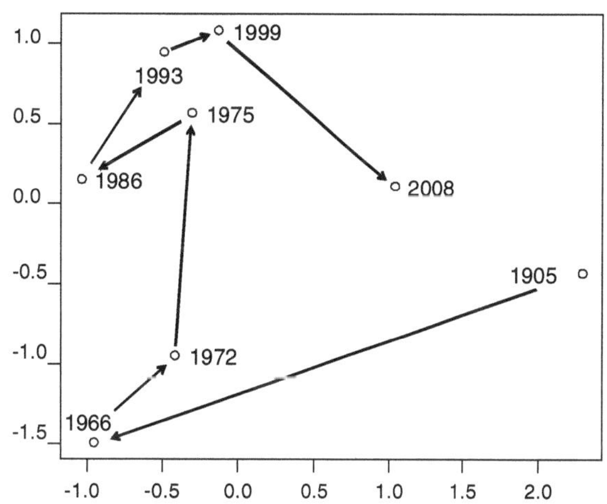


**Fig. 4** MDS plot of species occurrence the presence/absence data from all surveys. See text for details. *Arrows* illustrate apparent progression in aquatic plant community composition between surveys

lowest in 1966 (0.15) and highest in 2008 (0.53), and was generally constant between 1972 (0.37) and 1999 (0.34) (Fig. 1b).

At the more local scale, there appeared to be some increases in taxon richness. The taxa count per sample increased from a mean of about 1.5 in 1993 to about 2.2 in 2008 (Fig. 2). Similarly, the mean taxa count per survey sector increased from about 3.5 in 1993 to about 6 in 2008 (Fig. 3).

In the multi-dimensional scaling analysis (Fig. 4), 69% of variability in species composition was explained by the first two ordination axes with the first ($E = 9.7$, variability = 46%) explaining twice as much as the second ($E = 4.8$, variability = 23%). The first axis appears to be consistent with the diversity indicators generally tracking recovery from degradation, as suggested by the fact that the earliest modern samples were furthest away from 1905 and the 2008 survey was closest.

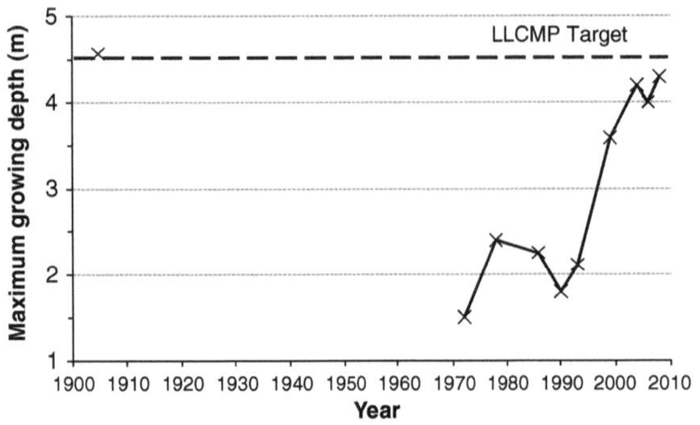


**Fig. 5** Maximum depth colonised by aquatic plants for all years in which the data are available. The *dashed line* represents the target for recovery set by the Loch Leven Catchment Management Plan (LLCMP, 1999; D'Arcy et al., 2006)

## Maximum growing depth

Maximum depth of colonisation, as reported by May & Carvalho (2010), declined from 1905 (4.6 m) to the early 1970 s (1.5 m), then showed some improvement in the late 1970s and 1980s (2.4 m in 1979). A dramatic improvement in this variable was recorded between 1990 (1.8 m) and the present day (4.3 m in 2008) (Fig. 5).

## Discussion

The aquatic plant community at Loch Leven has clearly undergone large changes since 1905. By all measures presented in this study, there was a decline between the early 1900s and the 1970/1980s. This was followed by a recovery that appeared to be continuing into 2008. This is consistent with similar patterns observed in water quality and other biota, as reported by Carvalho et al. (2011), Gunn et al. (2011), May et al. (2011) and Spears et al. (2011). Changes in species richness over time showed this pattern most simply; of the 18 taxa considered in this study, the highest number (16) occurred in 1905, while eight or less were found in the 1966, 1986 and 1993 surveys, and 13 were recorded in the most recent survey in 2008.

The increases in taxonomic evenness, richness per sample and richness per sector recorded over the last three comparable surveys seem to indicate that smaller-scale diversity (i.e. at the sector and sample scales) has been increasing, at least since the early 1990s. This supports the interpretation that increasing richness at the whole lake scale is consistent with improving water quality conditions. In particular, improvements in the late spring/early summer light climate (Carvalho et al., 2011) are likely to open up a much wider range of ecological niches within a transect or sector, in terms of both light requirements and substrate available for growth (Vestergaard & Sand-Jensen, 2000; Sand-Jensen et al., 2008). Increasing light availability may also protect plants from the limitations imposed by wave action and grazing by waterfowl (Jupp & Spence, 1977).

These improvements in community measures are supported by recent increases in maximum depth of colonisation by macrophyes. This measure was used as a target water quality indicator by the Loch Leven Catchment Management Plan (LLCMP, 1999), with the target being set to 4.5 m to reflect the value observed in 1905. This target appears to be almost realised, now.

Growing depths of 1.5 m in the early 1970s would have restricted substrate availability to the predominantly more wave-disturbed and sandier sediments (Jupp & Spence, 1977; Spears & Jones, 2010), whereas growing depths of up to 4.5 m allow additional species that are usually associated with more stable, finer sediments to flourish. In addition, the reductions in open water nutrient concentrations recorded (Carvalho et al., 2011), especially the lower soluble reactive phosphorus (SRP) concentrations for most months of the year and enhanced nitrogen limitation in summer, may reduce epiphyte burdens on macrophyte leaves and allow slower growing macrophytes, such as *Potamogeton praelongus*, to out compete faster growing species, such as *Potamogeton berchtoldii/pusillus* and *Chara globularis*.

In the absence of comparable quantitative data, the changes in survey methodologies imply a general increase in total aquatic plant biomass since 1966. Specifically, longer rake drags would have been used when submerged macrophytes were less abundant. In the earlier modern surveys (1966, 1972 and 1974), the sampling rake was allowed to drag along the bottom for 50–100 m (Jupp et al., 1974), a sampling strategy that would have been impossible in dense vegetation. In 1986, sampling distances of between 5 and 129 m were used (Robson, 1990). In all of the surveys since then, drag distances of only 2 m have been used (Murphy & Milligan, 1993; Griffin & Milligan, 1999;

this study). By the time of the most recent survey (2008), rake drags of longer than 5 m would have been very ineffective in retrieving a representative sample because the rake was often full after a 2 m drag, and any additional plant material would simply have failed to be collected.

It should be noted that taxon richness and other measures of diversity used in this study are sensitive to sampling effort (Wintle et al., 2004; Garrard et al., 2008). This is of particular concern for rare taxa. It is unfortunate that sampling effort cannot be defined precisely, here, due to the lack of information about methodologies for most of the survey data. However, it is clear that, if there is a bias, then it is generally in favour of the earlier modern surveys because, in these surveys, both the number of samples (Table 1) and the effort per sample (length of rake drag) were generally greater. This bias can only strengthen the conclusions made above regarding improvements in measures of aquatic plant diversity.

The results presented here clearly show that the Loch Leven macrophyte community is becoming more diverse in terms of species richness and evenness, and is becoming increasing similar to the species recorded in 1905. Palaeobotanical studies of aquatic plant macrofossils at Loch Leven (Salgado et al., 2009), however, suggest that even 1905 was not an undisturbed baseline. In the seventeenth and eighteenth century, even less competitive isoetids such as *Isoetes lacustris* and *Lobelia dortmanna*, were abundant in the loch and by 1905 conditions had probably already been altered by increasing levels of nutrient inputs (Salgado et al., 2009) and hydromorphological changes to the water level that were completed in 1831 (May & Spears, 2011). These modifications lowered the water level by 1.5 m (Morgan, 1970) and converted 2.7 $km^2$ of aquatic habitat to low quality farmland (Munro, 1994), raising the question of how to set appropriate recovery targets.

Two of the main reasons that Loch Leven is valued so highly in terms of its ecology are the internationally renowned over-wintering and breeding bird community (Carss et al., 2011) and the brown trout fishery (May & Spears, 2011; Winfield et al., 2011). There are indications that both of these communities, and the benthic invertebrates that form a key part of their diet, have responded positively to attempts to restore water quality in the loch (Carss et al., 2011; Gunn et al., 2011; Winfield et al., 2011). These broader recovery trends may be, in part, associated with the recovery observed in the macrophyte community, with the increased abundance and diversity of plants possibly providing greater physical habitat structure for invertebrates (van den Berg et al., 1997; Warfe & Barmuta, 2006), more productive and variable food sources for dabbling ducks and herbivorous birds such as coot and swan (Perrow et al., 1997; Moreno-Ostos et al., 2008) and more physical habitat for fish feeding and breeding (Warfe & Barmuta, 2006).

The comparison of multiple aquatic plant surveys has provided evidence that Loch Leven has been undergoing recovery from eutrophication since at least 1993. This is most evident when looking at species richness and at species evenness. The plant community has regained many, though not all, of the taxa that were present in 1905 and, as such, cannot be said to have returned to the state similar observed by West (1910) and probably never will. This is because, although nutrient inputs to the loch have been reduced by up to 60% (May et al., 2011), changes in land use and the engineering work that installed sluices on the loch's outlet (May & Spears, 2011) are unlikely ever to be reversed completely.

**Acknowledgments** The authors thank Bryan Spears and Linda May for help with the preparation of this manuscript. Thanks are also due to all of the surveyors involved in the collection of the data on which this article is based. We are also grateful to Kinross Estates for providing access to the loch. This research was funded by the Natural Environment Research Council. Loch Leven is part of the UK Environmental Change Network (http://www.ecn.ac.uk).

## References

Britton, R. H., 1975. Survey of aquatic macrophytes in Loch Leven. Loch Leven Research Group, Leven.

Carss, D., B. M. Spears, L. Quinn & R. Cooper, 2011. Long-term variations in waterfowl populations in Loch Leven: identifying discontinuities between local and national trends. Hydrobiologia. doi:10.1007/s10750-011-0927-6

Carvalho, L., C. Ferguson, B. M. Spears, I. D. M.Gunn, H. Bennion, A. Kirika & L. May, 2011. Water quality of Loch Leven: responses to enrichment, restoration and climate change. Hydrobiologia. doi:10.1007/s10750-011-0923-x.

D'Arcy, B. J., L. May, J. Long, I. R. Fozzard, S. Greig & A. Brachet, 2006. The restoration of Loch Leven, Scotland, UK. Water Science and Technology 53: 183–191.

Ferguson, C. A., L. Carvalho, E. M. Scott, A. W. Bowman & A. Kirika, 2008. Assessing ecological responses to environmental change using statistical models. Journal of Applied Ecology 45: 193–203.

Garrard, G. E., S. A. Bekessy, M. A. McCarthy & B. A. Wintle, 2008. When have we looked hard enough? A novel method for setting minimum survey effort protocols for flora surveys. Austral Ecology 33: 986–998.

Griffin, L. R. & A. Milligan, 1999. Submerged macrophytes of Loch Leven, Kinross (Report to Scottish Natural Heritage). University of Glasgow, Glasgow.

Gunn, I. D. M., M. T. O'Hare, P. S. Maitland & L. May, 2011. Long-term trends in Loch Leven invertebrate communities. Hydrobiologia. doi:10.1007/s10750-011-0926-7.

Jeppesen, E., M. A. Søndergaard, M. O. Søndergaard & K. Christofferson (eds), 1998. The structuring role of submerged macrophytes in lakes. Springer-Verlag, New York.

Jupp, B. P. & D. H. N. Spence, 1977. Limitations of macrophytes in a eutrophic lake, Loch Leven. II. Wave action, sediments and waterfowl grazing. Journal of Ecology 65: 431–446.

Jupp, B. P., D. H. N. Spence & R. H. Britton, 1974. Distribution and production of submerged macrophytes in Loch Leven, Kinross. Proceedings of the Royal Society of Edinburgh Section B-Biological Sciences 74: 195–208.

LLCMP, 1999. Loch Leven Catchment Management Plan. The Report of the Loch Leven Area Management Advisory Group: 93 pp.

Magurran, A. E., 2004. Measuring Biological Diversity. Blackwell, Oxford.

May, L. & L. Carvalho, 2010. Maximum growing depth of macrophytes in Loch Leven, Scotland, United Kingdom, in relation to historical changes in estimated phosphorus loading. Hydrobiologia 646: 123–131.

May, L. & B. M. Spears, 2011. Managing ecosystem services at Loch Leven, Scotland, UK: actions, impacts and unintended consequences. Hydrobiologia. doi:10.1007/s10750-011-0931-x

May, L. L. H. Defew, H. Bennion & A. Kirika, 2011. Historical changes (1905–2005) in external phosphorus loads to Loch Leven, Scotland, UK. Hydrobiologia. doi:10.1007/s10750-011-0922-y

Moreno-Ostos, E., M. Paracuellos, I. de Vicente, J. C. Nevado & L. Cruz-Pizarro, 2008. Response of waterbirds to alternating clear and turbid water phases in two shallow Mediterranean lakes. Aquatic Ecology 42: 701–706.

Morgan, N. C., 1970. Changes in the fauna and flora of a nutrient enriched lake. Hydrobiologia 35: 545–553.

Munro, D., 1994. Loch Leven and the River Leven–a landscape transformed. The River Leven Trust, Markinch: 196.

Murphy, K. J. & A. Milligan, 1993. Submerged macrophytes of Loch Leven, Kinross—August 1993. Report to Scottish Natural Heritage, Contract No. 937/F2A/5.3/224). Centre for Research in Environmental Science and Technology. University of Glasgow, Glasgow.

Murray, J. & L. Pullar, 1910. Bathymetrical survey of the freshwater lochs of Scotland, Edinburgh.

Penning, W. E., B. Dudley, M. Mjelde, S. Hellsten, J. Hanganu, A. Kolada, M. van den Berg, S. Poikane, G. Phillips, N. Willby & F. Ecke, 2008a. Using aquatic macrophyte community indices to define the ecological status of European lakes. Aquatic Ecology 42: 253–264.

Penning, W. E., M. Mjelde, B. Dudley, S. Hellsten, J. Hanganu, A. Kolada, M. van den Berg, S. Poikane, G. Phillips, N. Willby & F. Ecke, 2008b. Classifying aquatic macrophytes as indicators of eutrophication in European lakes. Aquatic Ecology 42: 237–251.

Perrow, M. R., J. H. Schutten, J. R. Howes, T. Holzer, F. J. Madgwick & A. J. D. Jowitt, 1997. Interactions between coot (*Fulica atra*) and submerged macrophytes: the role of birds in the restoration process. Hydrobiologia 342: 241–255.

R Development Core Team, 2008. R: a language and environment for statistical computing. R Foundation for Statistical Computing, Vienna.

Robson, T. O., 1986. Loch Leven, Kinross macrophyte survey—August 1986. Final report for Contract No HF3-03-208(46). Nature Conservancy Council.

Robson, T. O., 1990. Loch Leven—Kinross: macrophyte survey August 1990. Final Report to Nature Conservancy Council, Kinross. NCC Contract No. 21.F2B.219.DA01.

Salgado, J., S. Sayer, L. Carvalho, T. Davidson & I. Gunn, 2009. Assessing aquatic macrophyte community change through the integration of palaeolimnological and historical data at Loch Leven, Scotland. Journal of Paleolimnology 43: 191–204.

Sand-Jensen, K., N. L. Pedersen, I. Thorsgaard, B. Moeslund, J. Borum & K. P. Brodersen, 2008. 100 years of vegetation decline and recovery in Lake Fure, Denmark. Journal of Ecology 96: 260–271.

Schaumburg, J., C. Schranz, G. Hofmann, D. Stelzer, S. Schneider & U. Schmedtje, 2004. Macrophytes and phytobenthos as indicators of ecological status in German lakes—a contribution to the implementation of the Water Framework Directive. Limnologica 34: 302–314.

Smith, I. R., 1974. The structure and physical environment of Loch Leven, Scotland. Proceedings of the Royal Society of Edinburgh Section B-Biological Sciences 74: 81–100.

Søndergaard, M., E. Jeppesen, J. P. Jensen & S. L. Amsinck, 2005. Water framework directive: ecological classification of Danish lakes. Journal of Applied Ecology 42: 616–629.

Spears, B. M. & I. D. Jones, 2010. The long-term (1979–2005) effects of the North Atlantic oscillation index on wind-induced wave mixing in Loch Leven (Scotland). Hydrobiologia 646: 49–59.

Spears B. M., L. Carvalho, R. Perkins, A. Kirika & D. M. Paterson, 2011. Long-term variation and regulation of internal phosphorus loading in Loch Leven. Hydrobiologia. doi: 10.1007/s10750-011-0921-z

US Environmental Protection Agency, 1998. Lake and reservoir bioassessment and criteria: technical guidance document (No. Report EPA 841-B-98-007), Washington.

van den Berg, M. S., H. Coops, R. Noordhuis, J. van Schie & J. Simons, 1997. Macroinvertebrate communities in relation to submerged vegetation in two Chara-dominated lakes. Hydrobiologia 342: 143–150.

Vestergaard, O. & K. Sand-Jensen, 2000. Aquatic macrophyte richness in Danish lakes in relation to alkalinity, transparency, and lake area. Canadian Journal of Fisheries and Aquatic Sciences 57: 2022–2031.

Warfe, D. M. & L. A. Barmuta, 2006. Habitat structural complexity mediates food web dynamics in a freshwater macrophyte community. Oecologia 150: 141–154.

West, G., 1910. A further contribution to a comparative study of the dominant Phanerogamic and Higher Cryptogamic flora of aquatic habit in Scottish lakes. Proceedings of the Royal Society of Edinburgh, B, 30: 170–173, 162 Plates.

Winfield, I. J., C. E. Adams, R. Gardiner, A. Kirika, J. Montgomery, B. Spears, D. Stewart, J. E. Thorpe & W. Wilson, 2011. Changes in the fish community of Loch Leven: untangling anthropogenic pressures. Hydrobiologia. doi: 10.1007/s10750-011-0925-8.

Wintle, B. A., M. A. McCarthy, K. M. Parris & M. A. Burgman, 2004. Precision and bias of methods for estimating point survey detection probabilities. Ecological Applications 14: 703–712.

Hydrobiologia (2012) 681:59–72
DOI 10.1007/s10750-011-0926-7

LOCH LEVEN RESEARCH

# Long-term trends in Loch Leven invertebrate communities

I. D. M. Gunn · M. T. O'Hare · P. S. Maitland · L. May

Published online: 9 November 2011


**Abstract** Detailed studies of the macroinvertebrate benthos and zooplankton communities in Loch Leven, the largest shallow lowland lake in Scotland, UK, were carried out from 1966 to 1973 as part of the International Biological Programme (IBP). The results revealed a reduction in species diversity that was attributed to increasing eutrophication. This work provides a baseline against which the response of the invertebrate communities to subsequent changes in management can be assessed. This article compares macroinvertebrate benthos and zooplankton data from the IBP study with the post-IBP era during which changes at Loch Leven included a 60% reduction in the phosphorus input from external sources and variations in fish stocking rates. Only in recent years has there been evidence of ecological recovery by the invertebrate communities: the number of macroinvertebrate and zooplankton taxa has increased (including taxa considered to be sensitive to nutrient enrichment) and invertebrate abundances have declined. These changes appear to reflect the improvements in water quality and habitat conditions at Loch Leven that have occurred as a result of the recent reduction in nutrient loads, albeit with a substantial delay before any ecological response could be detected. This time lag in recovery has important implications for assessing improvements in the ecological status of other lake systems, as is required by the EU Water Framework Directive.

**Keywords** Eutrophication · Environmental Change Network · Biomonitoring · Lake · Benthos · Zooplankton · Recovery

Guest editors: L. May & B. M. Spears / Loch Leven: 40 years of scientific research

**Electronic supplementary material** The online version of this article (doi:10.1007/s10750-011-0926-7) contains supplementary material, which is available to authorized users.

I. D. M. Gunn (✉) · M. T. O'Hare · L. May
Centre for Ecology & Hydrology, Bush Estate, Penicuik, Midlothian EH26 0QB, UK
e-mail: idmg@ceh.ac.uk

P. S. Maitland
Fish Conservation Centre, Gladshot, Haddington, East Lothian EH41 4NR, UK

## Introduction

As a group, freshwater invertebrates are generally regarded as good indicators of the biological integrity of aquatic habitats and they play a key role in lake productivity, nutrient cycling and decomposition (Jones & Sayer, 2003; Jónasson, 2004; Brönmark & Hansson, 2005). It has been suggested that they have a wide range of environmental sensitivities and relatively rapid life cycles that can respond quickly to changes in lake processes (Williams & Feltmate, 1992; Rosenberg & Resh, 1993). In recognition of this, the EU Water Framework Directive (WFD) has included macroinvertebrate communities as one of the biological elements to be used for assessing the ecological status of lakes (European Parliament, 2000).

Historically, profundal and sub-littoral macroinvertebrates have been widely used as indicators of the trophic status of lake ecosystems (e.g. Cairns & Pratt, 1993; Brinkhurst, 2002), while the use of the littoral fauna for ecological monitoring has proven more challenging. Pressure and state relationships for littoral invertebrate communities are generally poorly understood (Håkanson, 2001) and they have a high temporal and spatial patchiness that may vary with lake type (White & Irvine, 2003). However, if sampling is limited to clearly defined meso-habitats, then littoral macroinvertebrates do seem to have the potential to be used to monitor lake responses to eutrophication and other pressures (O'Hare et al., 2007). Although zooplankton have been omitted from the WFD, many species are primary grazers, and thus their abundance and species composition are likely to be closely linked to changes in phytoplankton production and, as potential indicators of environmental conditions, they may also provide evidence of changes in the trophic status of lakes.

Within a lake, water quality and productivity change over time in response to variations in external drivers such as nutrient input and climate. A freshwater site that has been strongly affected by such changes over the last 40 years is Loch Leven, a large and shallow lake in Scotland, UK where long-term monitoring data has been collected since the late 1960s (Carvalho et al., 2011; Dudley et al., 2011; May & Spears, 2011; May et al., 2011;). Between 1966 and 1973, Loch Leven formed part of the International Biological Programme (IBP), a study of 43 lakes and 12 reservoirs located across a range of geographical regions from the tropics to the arctic, which aimed to take a whole ecosystem approach to understanding lake ecology (Brylinski & Mann, 1973).

The IBP project included studies of both of the macroinvertebrate benthos and zooplankton communities in Loch Leven (Royal Society of Edinburgh, 1974). By the late 1960s, the macroinvertebrate biomass had become dominated by Oligochaeta and Chironomidae. Given the pivotal role of Chironomidae in the benthic community as consumers of algal primary production, and as food for tertiary feeders such as the brown trout, the latter part of the IBP project focused on detailed studies of chironomid productivity (Maitland et al., 1972; Charles et al., 1974a, b, 1975; Maitland & Hudspith, 1974; Charles & East, 1977; Maitland, 1979). Overall, the benthos was found to be relatively poor in terms of species diversity compared with the late 19th century and the first half of the 20th century when Coleoptera, Ephemeroptera, Gastropoda and Trichoptera species were abundant (historical records summarised by Maitland & Hudspith, 1974). Similarly, the crustacean zooplankton had declined from 15 different species recorded at the end of the nineteenth century to only five species by the early 1970s (Johnson & Walker, 1974). This decline in species richness was attributed to the increasing eutrophication of the loch since the turn of the century (Morgan, 1970, 1974).

In recent years, attempts have been made to improve water quality at Loch Leven by reducing the inputs of nutrients from the catchment (e.g. Bailey-Watts & Kirika, 1987, 1999; LLCMP, 1999; May et al., 2011). Recent evidence suggests that this has resulted in an improvement in water quality (D'Arcy et al., 2006; Ferguson et al., 2008). Cyanobacterial blooms have reduced in frequency (Carvalho, pers comm.), spring water clarity has increased (Carvalho et al., 2011) and underwater plants have begun to thrive again in deeper water (Dudley et al., 2011; May & Carvalho, 2010). With the recovery of Loch Leven from eutrophication, comes a rare opportunity to investigate how the invertebrate communities have responded to this change using the IBP data (Royal Society of Edinburgh, 1974) as an historical baseline. Since the IBP project finished in the early 1970s, invertebrate monitoring, including that conducted under the auspices of the UK Environmental Change Network (ECN), has continued in Loch Leven (e.g., Fozzard, 1993; Lowery & Morrison, 1994, 1995, 1996; Morrison, 1997; Long, 1999, 2000), although, apart from work on rotifer ecology (May, 1980a, b, 1983, 1985, 1987a, b, 1989; May & Jones, 1989; May et al., 1992, 2001), most of this has not been published. Our study, therefore, aims to improve our understanding of the impact of eutrophication and recovery at Loch Leven on invertebrate communities and, more generally, to investigate the effectiveness of using invertebrates for the bio-monitoring of shallow lakes that are recovering from eutrophication. This issue is of increasing relevance given the current drive, under the WFD, to restore lakes to good ecological status.

## Materials and methods

### Study site

Loch Leven (56°10′N, 3°30′W) is a large, shallow, lowland, eutrophic lake, situated near Kinross in Scotland, UK. It lies in a predominantly agricultural catchment. The sediments of Loch Leven are composed of 42% sand, 57% mud and <1% stones (Maitland & Hudspith, 1974). The lake is described in detail in May & Spears (2011).

### Field sampling

#### *Zooplankton*

Crustacean zooplankton were sampled at weekly intervals between January 1975 and December 1982 at the Reed Bower and Centre Loch sampling stations (Fig. 1) and then not again until April 1989. Thereafter, samples were collected on an occasional basis until regular sampling was resumed in May 1992. Between 1975 and 1982, samples were collected either by vertical net haul (mesh size 100 μm) from 4 m depth to the surface, or with a tube sampler incorporating a filter of mesh size 125 μm (George & Owen, 1978) that was lowered to a depth of 4 to 5 m. From 1992 onwards, samples were collected by vertical net haul (mesh size 118 μm). In the laboratory, crustacean zooplankton samples were sub-sampled, and the animals were identified and counted.

Rotifer zooplankton samples were taken with a weighted plastic tube (Lund, 1949) at weekly intervals from 1977 to 1982 and, thereafter, sporadically collected with a bucket from just below the water surface throughout the rest of the 1980s. Regular sampling by bucket resumed from January 1989 and to June 1992, with most samples being taken at the Sluices sampling site (Fig. 1). Open water sampling (weekly to fortnightly) restarted in July 1992 and, since then, samples have been collected either using a 2-m plastic drainpipe (5 cm internal diameter) or an integrated tube sampler (3.00–3.75 m) at the Reed Bower sampling location, although samples collected from 1999 onwards have yet to be analysed. The rotifer samples were concentrated by repeatedly settling the samples in glass measuring cylinders and siphoning off the overlying water. The concentrated rotifer samples were counted with an inverted microscope (see May, 1980a, b, for details).

#### *Macroinvertebrate benthos*

The macroinvertebrate benthos in Loch Leven occurs in two main zones, the sub-littoral and the littoral. The mean depth of Loch Leven is 3.9 m and, although for much of the last 40 years mean Secchi depth transparency measurements have been less than this (Carvalho et al., 2011), the majority of the lake bed is potentially within the zone of effective light penetration, as demonstrated by macrophytes growing at depths in excess of 4 m (Dudley et al., 2011; May & Carvalho, 2010) when water quality is good. Following the definition of Solimini et al. (2006), i.e. that the sub-littoral is the bottom area of a lake that is covered by submerged macrophytes and algal vegetation, the majority of the Loch Leven 'open-water' benthos is classified as 'sub-littoral' rather than 'profundal'. The littoral zone, itself, is defined as the near-shore areas where emergent macrophytes grow (Solimini et al., 2006).

Summary data for the sub-littoral benthos from May and October 1970 was derived from chironomid production studies in the sandy areas undertaken by Maitland & Hudspith (1974). In addition, a spatially comprehensive 'snapshot' survey of 60 samples of the sub-littoral benthos (32 in sand and 28 in mud) of Loch Leven was carried out in May 1994. This survey used the same methodology (i.e. a Maitland corer over sand and a Jenkin surface sediment sampler over mud) as was used in 1968 (Maitland et al., 1972; Maitland, 1979).

The macroinvertebrate communities from three representative stony littoral sites, i.e. Castle Island, Kirkgate and either St Serf's Island (pre-2002) or the shore near the Sluices (post-2002) (Fig. 1), were sampled three times a year from 1998 to 2008, using a standard 3-min kick-sampling technique as part of the routine monitoring for the UK Environmental Change Network (ECN) (Sykes et al., 1999). Additional data from other surveys of the littoral benthos of the loch were compiled from the available literature (Maitland & Hudspith, 1974; Fozzard, 1993; Long, 1999, 2000).

### Data analysis

Recent and older data sets were compared, where possible, using statistical analyses. However, this approach could not be used for survey data that was available in summary form, only. Regression analyses

Fig. 1 Map of Loch Leven showing macroinvertebrate and zooplankton sampling sites

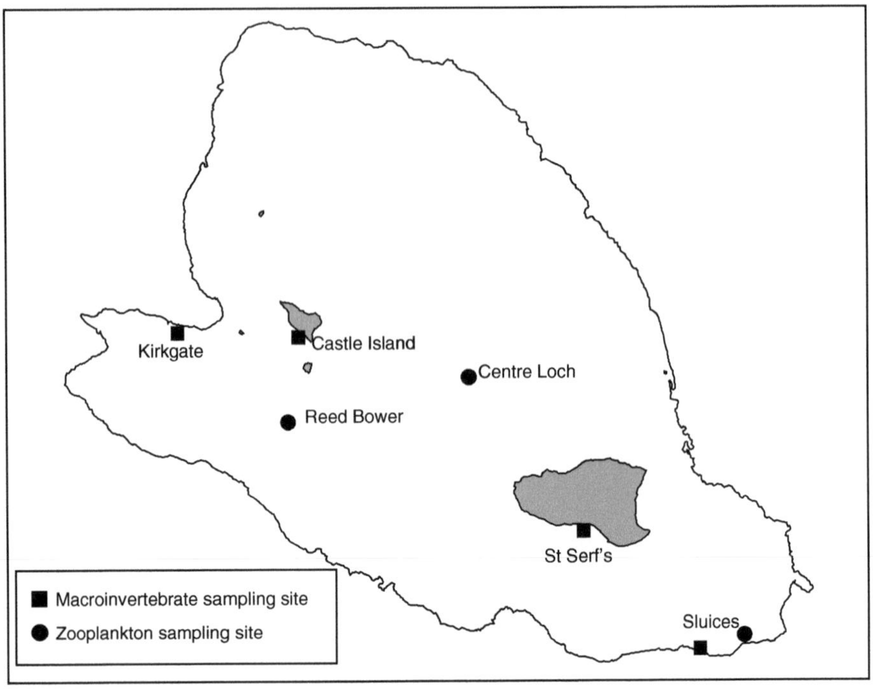


and a t-test were used to test for significant trends over time and for consistency of results across sites. In all cases, the data were tested for normality prior to analysis, using Anderson–Darling tests; it was found that no transformations were required. Minitab Version 13 was used for parametric analysis. Detrended Correspondence Analysis using the computer programme CANOCO (for Windows version 4.5) tested for seasonal, spatial or annual patterns in the ECN littoral benthos data.

## Results

The Loch Leven planktonic and littoral benthic invertebrate communities, monitored regularly since 1992 and 1998, respectively, have both undergone substantial changes since the IBP studies of the late 1960s and early 1970s. Both groups have become more species rich and show signs of a decline in total abundance.

### Crustacean zooplankton—species composition

The taxon richness of the crustacean zooplankton community between 1975 and 2007 was greater than in the IBP study that was undertaken in the late 1960s (Table 1). In this early study, the crustacean zooplankton was almost entirely composed of the copepod *Cyclops abyssorum*, although small numbers of *Bythotrephes longimanus*, *Leptodora kindti* and *Eudiaptomus gracilis* were also found (Johnson & Walker, 1974). In 1970, the *Daphnia hyalina* species complex was detected for the first time after an absence of 15–20 years (Johnson & Walker, 1974), the last published record being from August 1954 when it was found amongst the stomach contents of four trout (Morgan, 1970). From then onwards, the species composition of the crustacean zooplankton remained fairly stable with *Daphnia* and *C. abyssorum* co-dominating the community, the copepod *E. gracilis* occurring in smaller numbers, and the predatory cladocerans *L. kindti* and *B. longimanus* occurring occasionally and in very low numbers in summer (May et al., 1993; Gunn et al., 1994; Gunn & May, 1995, 1996, 1997, 1998, 1999). The community composition changed again in the 2000s, with *Cyclops vicinus* returning after a long period of absence (Carvalho et al., 2002, 2007, 2008) and *Bosmina longirostris* being recorded in large numbers for the first time in 2007 (Carvalho et al., 2008).

**Table 1** Crustacean zooplankton species in Loch Leven, late 1960s to 2007

| | Late 1960s | Early 1970s | 1990s | 2000s |
|---|---|---|---|---|
| Cladocera | | | | |
| *Daphnia hyalina* species complex | – | X | X | X |
| *Bosmina longirostris* (O.F.Müller) | – | – | – | X |
| *Bythotrephes longimanus* Leydig | X | X | X | X |
| *Leptodora kindti* (Focke) | X | X | X | X |
| Copepoda | | | | |
| *Eudiaptomus gracilis* (Sars) | X | X | X | X |
| *Cyclops abyssorum* Sars | X | X | X | X |
| *Cyclops vicinus* Uljanin | – | – | – | X |

Chydoridae and miscellaneous copepods associated with the sediments and aquatic vegetation, which are occasionally found in the plankton, are not included

## Crustacean zooplankton—abundance

Typically, the crustacean zooplankton in Loch Leven has been co-dominated by *Daphnia* and *Cyclops* populations since the early 1990s, with population maxima occurring in late spring/early summer followed by smaller secondary peaks in the autumn (e.g. Gunn & May, 1999). Carvalho et al. (2011) examined the data for evidence of long-term trends in *Daphnia* densities by comparing four seasonal population means with a range of water quality variables over the period 1975 to 2007, but found no trends to be significant. However, it became apparent that, over the last two decades, *Daphnia* densities had increased markedly in May (though not quite significantly) in response to warmer spring temperatures and that there was a general decline in *Daphnia* numbers between July and October. This is reflected in the statistically significant decline in the recorded mean summer (June to August) monthly maxima (Fig. 2, upper panel). *Daphnia* numbers build up in spring peaking in May and June. They then decrease but remain at higher densities throughout the rest of the summer than in the early months of spring. Overall, summer *Daphnia* numbers tend to reflect the spring phytoplankton bloom and continuing productivity during the summer period.

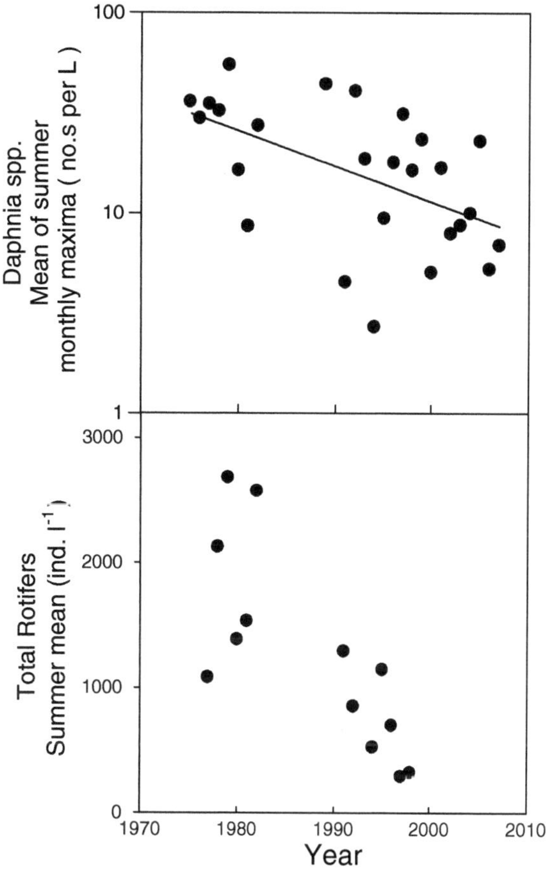


**Fig. 2** Trends in abundance of zooplankton in Loch Leven over the study period. Regression parameters for *Daphnia* are: $F = 9.25$, $n = 24$, $P < 0.01$, $R^2$ Adj. 24.8%. Summer mean rotifer numbers were significantly higher in the earlier sampling period (1977–1982) than in the later period (1991, 1992, 1994–1998). *t* test: *t* value of 3.79, df = 7, $P < 0.01$)

## Rotifer zooplankton—species composition

Nine rotifer species were recorded from samples collected between 1969 and 1970 (Johnson & Walker, 1974), whilst samples collected between 1977 and 1982 contained 14 species (May, 1980a) and those collected between 1991 and 1998, 21 species (Table 2). Samples collected between 1977 and 1982 and between 1991 and 1998 are directly comparable, having been collected over a similar period and with comparable sampling effort. When compared, it was found that all of the species present in the earlier period were also recorded in the later period. However, a 50% increase in species was noted in the latter period, with new records for: *Collotheca mutabilis*,

**Table 2** Planktonic rotifer species in Loch Leven, late 1960s to 1998

| Species | 1969–1970 | 1977–1982 | 1991–1998 |
|---|---|---|---|
| *Asplanchna priodonta* Gosse 1850 | X | X | X |
| *Brachionus angularis* Gosse 1851 | – | X | X |
| *Brachionus rubens* (Ehrenberg) | X | – | |
| *Collotheca mutabilis* (Hudson 1885) | – | – | X |
| *Colurella adriatica* Ehrenberg 1831 | – | X | X |
| *Conochilus hippocrepis* (Schrank 1830) | – | – | X |
| *Conochilus unicornis* Rousselet 1892 | X | X | X |
| *Filinia longiseta* (Ehrenberg 1834) | X | X | X |
| *Kellicottia longispina* (Kellicott 1879) | – | – | X |
| *Keratella cochlearis* (Gosse 1851) | X | X | X |
| *Keratella quadrata* (Müller 1786) | X | X | X |
| *Notholca squamula* Müller 1786 | – | X | X |
| *Polyarthra dolichoptera* Idelson 1925 | X | X | X |
| *Polyarthra euryptera* (Wierzejski 1893) | – | X | X |
| *Polyarthra major* (Burckhardt 1900) | – | – | X |
| *Polyarthra vulgaris* (Carlin 1943) | – | – | X |
| *Pompholyx sulcata* Hudson 1885 | X | X | X |
| *Synchaeta grandis* Zacharias 1893 | – | X | X |
| *Synchaeta kitina* Rousselet 1892 | – | X | X |
| *Synchaeta oblonga* (Ehrenberg 1832) | – | – | X |
| *Synchaeta pectinata* Ehrenberg | X | – | |
| *Synchaeta tremula* (Müller 1786) | – | – | X |
| *Trichocerca pusilla* Lauterborn 1898 | – | X | X |

*Conochilus hippocrepis*, *Kellicottia longispina*, *Polyarthra vulgaris*, *Synchaeta oblonga*, *Synchaeta pectinata* and *Synchaeta tremula*. Although most of these additional species can be found in many types of lake across the trophic gradient, it is interesting to note that *C. hippocrepis* and *K. longispina* are usually associated with oligotrophic systems (Ruttner-Kolisko, 1974), suggesting an improvement in water quality in the 1990s compared with 1977–1982. The appearance of *S. tremula* in the 1990s probably reflects an increase in macrophyte coverage (Dudley et al., 2011; May & Carvalho, 2010), as this species tends to live in the littoral zone (Ruttner-Kolisko, 1974).

## Rotifer zooplankton—abundance

Total rotifer abundance was relatively low during the 1990s compared with 1977–1982, *t* test, *t* value 3.79, df $= 7$, $P < 0.01$ (Fig. 2, lower panel). This suggests that the reduction in nutrient input to the loch, which began in the late 1980s, strongly influenced rotifer abundance, probably through its impact on food availability. Gunn & May (1997, 1998, 1999) examined the relationship between *Daphnia* numbers and total rotifer abundance in Loch Leven over several years in the 1990s and found that short-term variations in total rotifer abundance were inversely related to *Daphnia* densities. It was concluded that, when *Daphnia* numbers were low, rotifer abundance increased due to decreased competition for their preferred phytoplankton food. It is unlikely that this effect was caused by interference competition (i.e. from physical damage to the rotifer population from *Daphnia* filtering), as has been suggested for other sites (e.g. Burns & Gilbert, 1986), because experimental studies showed that this does not affect the population dynamics of rotifers and *Daphnia* in Loch Leven (May & Jones, 1989).

## Littoral benthos—species composition

The littoral community is now dominated numerically by Crustacea (*Asellus aquaticus* and *Gammarus pulex*), Oligochaeta (Tubificidae & Lumbriculidae),

Chironomidae, Trichoptera (*Tinodes waeneri*, *Agraylea multipunctata*) and Ephemeroptera (*Caenis luctuosa*). The full list of macroinvertebrate species recorded from Loch Leven's littoral zone is given in Appendix 1 (Supplementary material). Changes in the total number of macroinvertebrate taxa recorded in the littoral area of Loch Leven over the study period are shown in Fig. 3. Ninety-six taxa were recorded pre-1966, 41 taxa between 1966 and 1973 (IBP studies), and 153 taxa between 1993 and 2006. Of these, 24 of the taxa recorded pre-1966 were not recorded again and 48 of the taxa recorded in 1993–2006 were recorded for the first time. All of the taxa collected in 1966–1973 were also found on, at least, one more occasion. As sampling effort and collection method differed among the studies, the opportunities for direct comparisons of these records are limited. However, it is noteworthy that the numbers of Plecoptera, Ephemeroptera, Coleoptera and Trichoptera species recorded during the ECN monitoring (1998–2006) were much higher than those recorded in earlier surveys (Fig. 4). In the later surveys, the species recorded included pollution intolerant species such as *Diura bicaudata* and *Ecdyonurus dispar*.

## Littoral benthos—abundance/species richness at ECN sites

There was little change in relative total sample abundance or species richness, over time, either at Castle Island or Kirkgate (Fig. 5). However, from 2006 to 2008, abundance and species richness (all seasons combined) were lower than previously recorded at both of these ECN sites.

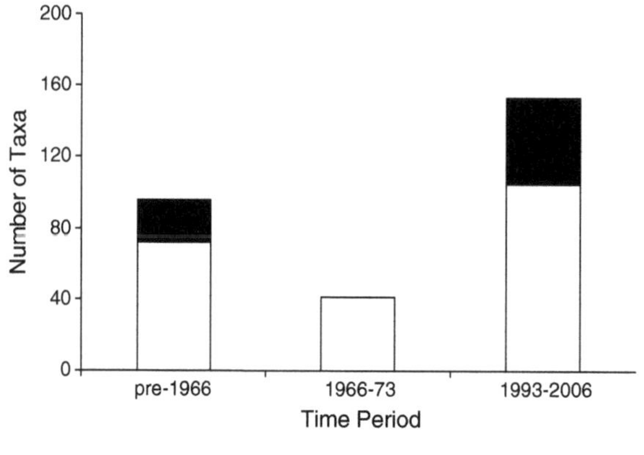


**Fig. 3** Total number of macroinvertebrate taxa recorded in studies of the littoral area of Loch Leven over time; the *black shaded areas* show the number of taxa that were unique to one time period, only

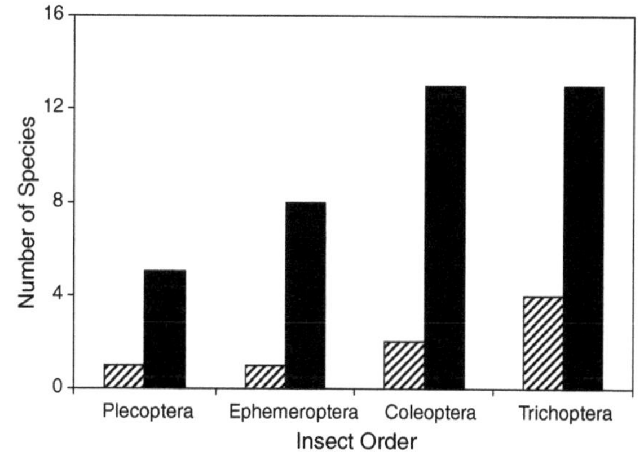


**Fig. 4** The number of species in selected major insect groups recorded during the IBP (1966–1973; *cross hatching*) and ECN (1998–2006; *black shading*) time periods; the pre-1973 data are taken from (Maitland & Hudspith, 1974)

## Littoral benthos—spatial and temporal patterns at ECN sites

The ordination analysis revealed that the biggest difference in community composition within the ECN samples was not related to location or year, but the season in which the samples were collected. Spring samples were distinct from summer samples, with autumn values overlapping to some extent with both spring and summer (Fig. 6).

## Sub-littoral benthos—species composition and distribution

The species composition, mean densities and spatial distribution of the sub-littoral benthos recorded in May 1994 (Table 3; Appendix 2 in Supplementary material) were broadly similar to that found in the IBP studies (Maitland & Hudspith, 1974; Charles et al., 1974a). The most abundant macroinvertebrate groups were Nematoda, Oligochaeta (mainly Tubificidae), Diptera (mainly Chironomidae) and Mollusca (mainly *Pisidium* spp. and *Valvata piscinalis*). New species records included a chironomid genus, *Paracladopelma*, considered to be a genera relatively tolerant of organic pollution (Wilson & Ruse, 2005), and three oligochaete species: the naid *Uncinais ucinata* and the tubificids *Psammoryctides barbatus* and *Limnodrilus*

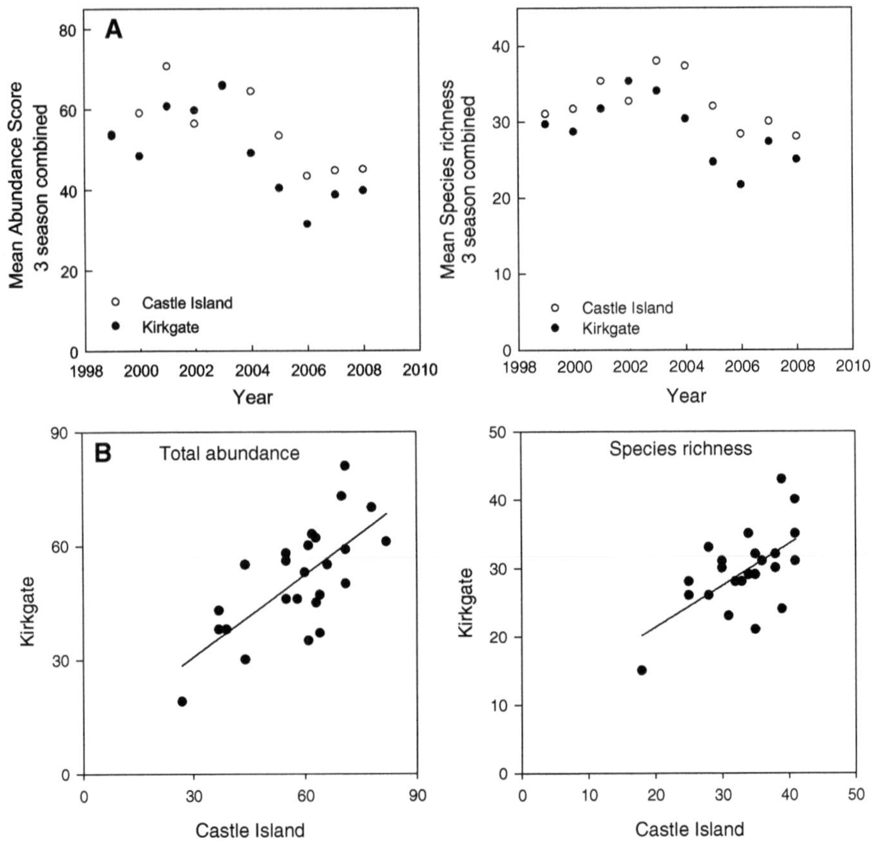


**Fig. 5** Comparison of metrics calculated from macroinvertebrate data collected at two Environmental Change Network (ECN) littoral sites, Kirkgate and Castle Island, in Loch Leven between 1998 and 2008

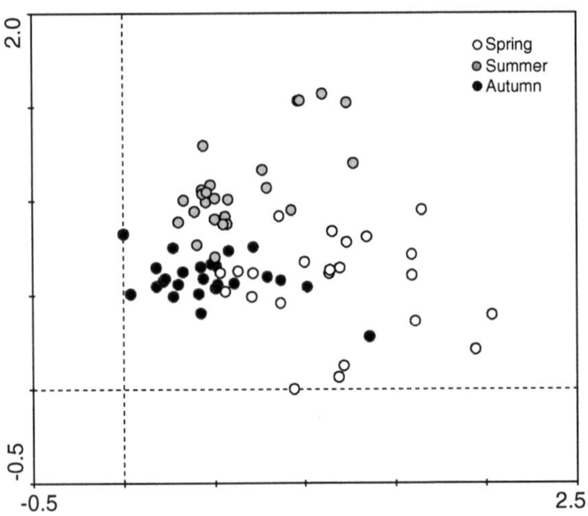


**Fig. 6** Detrended Correspondence Analysis (DCA) of the three season Environmental Change Network (ECN) macroinvertebrate data from 1998 to 2006

*claparedeianus*, which have also been recorded in other eutrophic lakes in the UK (Potter & Learner, 1974; Carter & Murphy, 1993). The chironomid *Endochironomus,* a taxa considered tolerant of organic pollution (Wilson & Ruse, 2005), which had not been recorded post-1969 in the IBP chironomid production studies (Charles et al., 1974a), re-appeared in 1994. A number of taxa found during the IBP studies were not recorded in the May 1994 survey. These included a number of chironomid genera including *Diamesa, Microtendipes, Harnischia, Micropsectra, Pentaneura, Psilotanypus* and several naid species, i.e. *Stylaria lacustris, Nais paradalis* and *Nais variables.*

## Sub-littoral benthos—abundance (1968, 1970 vs. 1994)

Summary abundance data were available for the major groups of sub-littoral benthos recorded in the sandy areas for some of 1968 and 1970 (Maitland & Hudspith, 1974). These were compared to the May 1994 survey data. Total abundance was substantially higher in May 1994 than in May 1970. *Pisidium*, Oligochaeta and Nematoda were all more abundant in 1994 than in the earlier years, although Chironomidae numbers were lower (Table 3). The Oligochaete to

**Table 3** Density (individuals $m^{-2}$) of the major taxonomic groups of macroinvertebrates found in the sand area of Loch Leven in 1968, 1970 and 1994; the ratio of Oligochaeta to Chironomidae is also shown

| Macroinvertebrate taxonomic group | October | | May | |
|---|---|---|---|---|
| | 1968 | 1970 | 1970 | 1994 |
| Hirudinea | 158 | 109 | 29 | 93 |
| Gastropoda | 255 | 206 | 177 | 77 |
| Pisidium | 995 | 840 | 199 | 1,635 |
| Chironomidae | 21,411 | 23,039 | 16,784 | 14,414 |
| Oligochaeta | 25,934 | 3,729 | 3,330 | 6,876 |
| Nematoda | 34,953 | 243 | 227 | 8,176 |
| Total | 83,706 | 28,166 | 20,746 | 31,271 |
| Oligochaeta/ Chironomidae ratio | 1.21 | 0.16 | 0.20 | 0.48 |

Chironomid ratio, often used in applied limnology as an indicator of trophic conditions, with higher ratios indicating increased enrichment (e.g. Saether, 1979), was also higher in 1994 than during an equivalent period in 1970. This suggested that water quality was poorer in 1994 than 1970, although better than that recorded for October 1968.

## Discussion

Loch Leven is a shallow lake in central Scotland, UK that has a well-documented history of increasing and then decreasing nutrient inputs over the last 100 years (May et al., 2011). When exceptionally high phosphorus inputs to the loch were recorded in 1985, a programme of nutrient reduction was put in place through targeted catchment management (LLCMP, 1999). This resulted in a 60% decrease in the phosphorus (P) input to the loch between 1985 and 1995 (May et al., 2011). Although recycling of P from the sediments prevented an immediate improvement in water quality (Carvalho et al., 2011; Spears et al., 2011), there are clear signs that the water quality is now improving (Carvalho et al., 2011), with macrophytes returning to deeper water (Dudley et al., 2011; May & Carvalho, 2010). Between the periods 1968–1977 and 1998–2007 the average annual means of total phosphorus (TP) and chlorophyll *a* concentrations declined in Loch Leven from 81 and 60 $\mu g\ l^{-1}$ to 59 and 37 $\mu g\ l^{-1}$, respectively (Carvalho et al., 2011). Carvalho et al. (2011) also showed, that over this study period, there was a highly significant, increasing linear trend in spring air temperatures.

Lakes that are undergoing eutrophication are normally characterised by increases in nutrient concentrations and phytoplankton production (often including increased frequencies and intensities of toxic cyanobacterial blooms). These changes have the potential to reduce water clarity and increase the supply of organic matter to the sediments which, in turn, may lower oxygen availability and reduce structural complexity within the benthic zone resulting in an impact on phytobenthos production (Heiskanen & Solimini, 2005). Metrics of the likely ecological impacts of eutrophication on benthic and pelagic invertebrate communities are increased absolute levels of abundance (reflecting the increased productivity of nutrient-enriched systems) and reduced species diversity (indicating a reduction in suitable physiological conditions for many species to thrive in).

Detailed sampling of the invertebrate community only commenced when Loch Leven was already impacted, making it difficult to set targets and assess recovery. However, evidence for their being a time lag in the expected recovery of some components of the fauna, after the implementation of pollution controls enacted in the late 1980s, is provided by the results of the 1994 "snap-shot" study of the sub-littoral benthos. It showed that the fauna was still indicative of eutrophic conditions and was comparable with that found in the IBP studies in the 1960s and early 1970s. Although the rest of the Loch Leven invertebrate dataset is patchy post-IBP with, for example, some large gaps in the zooplankton data, some encouraging trends pointing towards ecological recovery from eutrophication are discernible. In the years after the restoration programme began, species diversity (including species known to be less tolerant of eutrophication) increased in both the zooplankton and the littoral benthos communities and total abundance declined, especially in the rotifer zooplankton community.

It should be noted, however, that the most significant change in the crustacean zooplankton community occurred in 1970, when *Daphnia* was detected after a temporary absence of some 15–20 years. This occurred before the restoration programme began and is believed to have been a response to reductions in the discharge of dieldrin by a local mill (D'Arcy et al., 2006; May & Spears, 2011) rather than to any change in nutrient input. The re-appearance of *Daphnia* had a major impact on zooplankton (Johnson & Walker, 1974) and phytoplankton communities (Bailey-Watts, 1974) in the early 1970s as well as on the macroinvertebrate benthos, with the latter shifting from detrital/phytoplankton feeding species to benthic algal feeding species (Johnson & Walker, 1974; Maitland & Hudspith, 1974). Carvalho et al. (2011) carried out a detailed examination of trends in *Daphnia* numbers at Loch Leven in relation to changing water quality and concluded that, in the early 1970s, TP and chlorophyll *a* concentrations dropped abruptly as a result of the re-appearance of *Daphnia* in the lake. However, since 1988 the lower densities of *Daphnia* observed in summer cannot be explained by phytoplankton biomass, as summer chlorophyll *a* trends show no significant decline over the same period. Carvalho et al. (2011) suggest that the decline in mean and maxima *Daphnia* densities over the last decade may, therefore, be as a result of either an increased proportion of poorer quality food (e.g. inedible or toxic algae) in the *Daphnia* diet or increased predation from zooplanktivorous fish.

Fish predation on *Daphnia* in Loch Leven may have increased because due to stocking with brown trout (*Salmo trutta*, L.) and rainbow trout (*Oncorhynchus mykiss* Walbaum) in 1983–2006 and 1993–2004, respectively, to boost angling catches (Duncan, 1994; May & Spears, 2011; Winfield et al., 2011). The proportion of macroinvertebrates and zooplankton in the diets of brown trout (*Salmo trutta* L.) collected between 1993 and 1995 (Lowery & Morrison, 1994, 1995, 1996; Morrison, 1997) has been compared to Nature Conservancy data collected over the period 1967–1970 (Maitland, unpublished data). This analysis showed that, by the 1990s, chironomid larvae, although still the dominant food item, had become less important as a food source, with Mollusca and Crustacea (including zooplanktonic microcrustacea) making up a greater proportion of the diet. Although detailed information on prey availability is not available for the two study periods, it is known that chironomids were present in Loch Leven in high numbers in the late 1960s, which may explain their dominance in the diet of trout at the time. In a separate study, Yang et al. (1999) showed that the densities and size frequency of the *Daphnia* population in Loch Leven changed little over the period 1992–1995, suggesting that stocking with rainbow trout which began in 1993 was less important than other factors, such as food resources and water temperature, in controlling *Daphnia* numbers at that time. Overall, improvements in water quality seem to have resulted in a general decline in crustacean zooplankton abundance coupled with a small increase in species diversity over the period of restoration. Changes in species composition and total abundance were more marked in the rotifer community than in the crustacean zooplankton community, suggesting that rotifers respond more readily to changes in nutrient inputs and, subsequently, water quality than their crustacean counterparts. Further analysis of the rotifer samples collected from 1999 onwards would be instructive in seeing whether this trend of declining total abundance and increased species diversity has continued into the last decade.

Classification and monitoring schemes based on macroinvertebrates have, traditionally, focused on the profundal fauna of deep lakes (e.g. Saether, 1979; Wiederholm, 1981), where more uniform environmental conditions prevail. In general, macroinvertebrates from the littoral areas of lakes such as Loch Leven, where a sub-littoral zone predominates, have received relatively little attention as potential environmental indicators. Solimini et al. (2006) hypothesised that benthic invertebrate communities in this zone would be less affected by eutrophication than the benthos in the profundal region and Moss et al. (2003) concluded that metric indicators based on functional status might be the most reliable way of assessing ecological condition in this area. Such metrics include the ratio of oligochaetes to chironomids (O:C) in soft sediment samples. The data from Loch Leven seem to support the use of the O:C ratio and the sub-littoral benthos, in general, as a measure of environmental conditions in these areas, and is in agreement with similar findings from other systems (e.g. Lang & Reymond, 1992).

Solimini et al. (2006) speculated further that littoral invertebrate communities would be affected even less by eutrophication than the sub-littoral benthos. It has been suggested that the heterogeneity of shallow lake habitats causes differences within the benthos that would mask any changes associated with changes in nutrient status (Moss et al., 2003). However, because the ECN littoral benthos sampling method used in this study is limited to a single type of shoreline (i.e. stony) and the same sites are used each year, the differences

**Table 4** Summary of biotic signals, by invertebrate group, indicative of ecological recovery in Loch Leven since the IBP baseline studies of 1966–1973

| Invertebrate group | Species composition | Abundance |
|---|---|---|
| Crustacean zooplankton | Increased species diversity | Increased *Daphnia* densities in May; lower mean summer monthly maxima |
| Rotifer zooplankton | Increased species diversity | Total rotifer abundance significantly declined in 1990s |
| Littoral benthos | Increased species diversity | Unclear—insufficient data to draw conclusions |
| Sub-littoral benthos | Unclear—no data since 1994 | Unclear—no data since 1994 |

caused by habitat heterogeneity should be minimised and sensitivity to other types of change, such as lake wide eutrophication, maximised (White & Irvine, 2003; Brauns et al., 2007; O'Hare et al., 2007). The value of the Loch Leven system is that it is one of only a few lakes that have long-term records over a period of recovery from eutrophication with which this hypothesis can be tested. The results of this study only partially support these assertions. Regression analysis of metrics recorded at two separate sampling sites did show that different locations can produce comparable results year-on-year for eutrophication sensitive metrics such as total abundance. However, temporal variation in species richness differed at the two sites. This may have been caused by changes in local conditions, such as variations in water level, which can expose different combinations of meso-habitat at different sites to sampling. Hence, macro-invertebrates at individual sites may respond more to changes in local, rather than lake-wide, conditions in some cases. Therefore, care should be taken when interpreting littoral invertebrate data as it may not be indicative of water quality, alone.

## Conclusions

All of the available macroinvertebrate and zooplankton data collected in Loch Leven since the baseline IBP study in the late 1960s/early 1970s have been compiled. Results from detailed analysis of the trends in all of the different components of the invertebrate community, despite large gaps in the data, have shown, overall, a positive response, in terms of increased species diversity and reduction in total abundance (Table 4). There was, however, a substantial delay in the ecological recovery of some elements of the fauna in response to changes in management practices and reductions in nutrient input at Loch Leven. This time lag has important implications for assessing improvements in the ecological status of other lake systems under the EU Water Framework Directive.

**Acknowledgments** We thank Alex Kirika for his dedicated sampling of Loch Leven over many years. We are grateful to Scottish Natural Heritage (SNH) and the Loch Leven Fisheries (especially Willie Wilson) for practical and logistical help with field sampling. We also thank Glen George and David Jones for maintaining the crustacean zooplankton monitoring during the late 1970s and early 1980s. The supply of macroinvertebrate data by the Scottish Environment Protection Agency (SEPA) is gratefully acknowledged. The constructive comments of the referees greatly improved this manuscript. Macroinvertebrate benthos and zooplankton monitoring has been supported by the Natural Environment Research Council (NERC), SNH and SEPA (formerly, the Forth River Purification Board). Loch Leven is part of the UK Environmental Change Network ( http://www.ecn.ac.uk/).

## References

Bailey-Watts, A. E., 1974. The algal plankton of Loch Leven, Kinross. Proceedings of the Royal Society of Edinburgh, B 74: 135–156.

Bailey-Watts, A. E. & A. Kirika, 1987. A re-assessment of the phosphorus inputs to Loch Leven (Kinross, Scotland): rationale and an overview of results on instantaneous loadings with special reference to runoff. Transactions of the Royal Society of Edinburgh, Earth Sciences 78: 351–367.

Bailey-Watts, A. E. & A. Kirika, 1999. Poor water quality in Loch Leven (Scotland) in 1995, in spite of reduced phosphorus loadings since 1985: the influences of catchment management and inter-annual weather variation. Hydrobiologia 403: 135–151.

Brauns, M., X. F. Garcia, M. T. Pausch & N. Walz, 2007. Eulittoral communities of lowland lakes: discrimination among trophic states. Freshwater Biology 52: 1022–1032.

Brinkhurst, R. O., 2002. The Benthos of Lakes. The Blackburn Press, Caldwell, NJ.

Brönmark, C. & L.-A. Hansson, 2005. The Biology of Lakes and Ponds. University Press, Oxford.

Brylinski, M. & K. H. Mann, 1973. An analysis of factors governing productivity in lakes and reservoirs. Limnology and Oceanography 18: 1–14.

Burns, C. W. & J. J. Gilbert, 1986. Effects of daphnid size and density on interference between *Daphnia* and *Keratella cochlearis*. Limnology and Oceanography 31: 848–858.

Cairns, J. & J. R. Pratt, 1993. A history of biological monitoring using benthic macroinvertebrates. In Rosenberg, D. M. & V. H. Resh (eds), Freshwater Biomonitoring and Benthic Macroinvertebrates. Chapman & Hall, London.

Carter, C. E. & P. M. Murphy, 1993. The macroinvertebrate fauna of Lough Neagh. In Wood, R. B. & R. V. Smith (eds), Lough Neagh: The ecology of a multipurpose resource. Kluwer Academic Publishers, Dordrecht.

Carvalho, L., A. Kirika & I. Gunn, 2002. Long-term Patterns of Change in Physical, Chemical and Biological Aspects of Water Quality at Loch Leven. Unpublished report to Scottish Natural Heritage, Centre for Ecology & Hydrology.

Carvalho, L., B. Dudley, I. Gunn & A. Kirika, 2007. Loch Leven 2004–2006: trends in water quality and biological communities. Report to Scottish Natural Heritage, Centre for Ecology & Hydrology.

Carvalho, L., B. Spears, B. Dudley, I. Gunn, S. Zimmerman, L. Defew & L. May, 2008. Loch Leven 2007: trends in water quality and biological communities. Report to the Scottish Environment Protection Agency, Centre for Ecology & Hydrology.

Carvalho, L., C. Ferguson, B. M. Spears, I. D. M. Gunn, H. Bennion, A. Kirika & L. May, 2011. Water quality of Loch Leven: responses to enrichment, restoration and climate change. Hydrobiologia. doi:10.1007/s10750-011-0923-x.

Charles, W. N. & K. East, 1977. Changes in the structure of some mouth parts between first and second instar larvae in four species of Chironominae (Diptera, Chiromonidae). Entomologist's Gazette 28: 285–287.

Charles, W. N., K. East, D. Brown, M. C. Gray & T. D. Murray, 1974a. The production of larval Chironmidae in the mud at Loch Leven, Kinross. Proceedings of the Royal Society of Edinburgh, B 74: 241–258.

Charles, W. N., K. East, D. Brown, M. C. Gray & T. D. Murray, 1974b. Production studies on the larvae of four species of Chironomidae in the mud at Loch Leven in Scotland during 1970–71. Entomologisk Tidskrift 95: 34–41.

Charles, W. N., K. East & T. D. Murray, 1975. Production of larval Tanypodinae (Insecta: Chironmidae) in the mud at Loch Leven, Kinross. Proceedings of the Royal Society of Edinburgh, B 75: 157–169.

D'Arcy, B. J., L. May, J. Long, I. R. Fozzard, S. Greig & A. Brachet, 2006. The restoration of Loch Leven, Scotland. Water Pollution Research 53: 183–191.

Dudley, B., I. D. M. Gunn, L. Carvalho, I. Proctor, M. T. O'Hare, K. J. Murphy, A. Milligan, 2011. Changes in aquatic macrophyte communities in Loch Leven—evidence of recovery from eutrophication? Hydrobiologia. doi:10.1007/s10750-011-0924-9.

Duncan, W., 1994. The fish community in Loch Leven. In Hutchinson, P. & A. Walker (eds), The Loch Leven Trout Fishery–Its Future. Institute of Fisheries Management, Pitlochry.

European Parliament, 2000. Directive 2000/60/EC of the European Parliament and of the Council establishing a framework for community action in the field of water policy. Official Journal of the European Communities L 327: 1–72.

Ferguson, C. A., L. Carvalho, E. M. Scott, A. W. Bowman & A. Kirika, 2008. Assessing ecological responses to environmental change using statistical models. Journal of Applied Ecology 41: 193–203.

Fozzard, I., 1993. Loch Leven macroinvertebrate survey, 20th April and 31st August 1993. Internal Report, Forth River Purification Board.

George, D. G. & G. H. Owen, 1978. A new tube sampler for crustacean zooplankton. Limnology and Oceanography 23: 563–566.

Gunn, I. D. M. & L. May, 1995. Analysis of 1994 Loch Leven Zooplankton samples. Report to Scottish Natural Heritage, Institute of Freshwater Ecology.

Gunn, I. D. M. & L. May, 1996. Analysis of zooplankton samples 1995—Loch Leven NNR. Report to Scottish Natural Heritage, Institute of Freshwater Ecology.

Gunn, I. D. M. & L. May, 1997. Analysis of 1996 zooplankton samples—Loch Leven NNR. Report to Scottish Natural Heritage, Institute of Freshwater Ecology.

Gunn, I. D. M. & L. May, 1998. Analysis of 1997 zooplankton samples—Loch Leven NNR. Report to Scottish Natural Heritage, Institute of Freshwater Ecology.

Gunn, I. D. M. & L. May, 1999. Analysis of 1998 zooplankton samples—Loch Leven. Report to Scottish Natural Heritage, Institute of Freshwater Ecology.

Gunn, I. D. M., L. May & A. E. Bailey-Watts, 1994. Analysis of 1993 Loch Leven zooplankton samples. Report to Scottish Natural Heritage, Institute of Freshwater Ecology.

Håkanson, L., 2001. Assessment of critical loading of lakes as a basis for remedial measures: a review of fundamental concepts. Lakes and Reservoirs: Research and Management 6: 1–20.

Heiskanen, A.-S. & A. G. Solimini, 2005. Relationships between pressures, chemical status, and biological quality elements: Analysis of the current knowledge gaps for the implementation of the Water Framework Directive. Office for Official Publications of the European Communities, Luxembourg.

Johnson, D. & A. F. Walker, 1974. The zooplankton of Loch Leven, Kinross. Proceedings of the Royal Society of Edinburgh, B 74: 285–294.

Jónasson, P. M., 2004. Benthic invertebrates. In O'Sullivan, P. E. & C. S. Reynolds (eds), The Lakes Handbook, Vol. 1. Blackwell Science Ltd, Malden: 341–416.

Jones, J. I. & C. D. Sayer, 2003. Does the fish-invertebrate-periphyton cascade precipitate plant loss in shallow lakes? Ecology 84: 2155–2167.

Lang, C. & O. Reymond, 1992. Reversal of eutrophication of Lake Geneva: evidence from the oligochaete communities. Freshwater Biology 28: 145–148.

LLCMP, 1999. Loch Leven Catchment Management Plan: The Report of the Loch Leven Area Management Advisory Group: 93 pp.

Long, J., 1999. Invertebrate survey of macrophyte beds, Loch Leven, 1998. Internal report, Scottish Environment Protection Agency.

Long, J. 2000. Invertebrate survey of macrophyte beds, Loch Leven, 1999. Internal report, Scottish Environment Protection Agency.
Lowery, G. & B. Morrison, 1994. Analysis of the Stomach Contents of Brown Trout, Rainbow Trout and Perch from Loch Leven, Fife Unpublished report to Scottish Natural Heritage. Freshwater Fisheries Laboratory, Pitlochry.
Lowery, G. & B. Morrison, 1995. Brown Trout and Rainbow Trout Diet Analysis, Loch Leven NNR 1994. Report to Scottish Natural Heritage, Freshwater Fisheries Laboratory, Pitlochry.
Lowery, G. & B. Morrison, 1996. Analysis of the Stomach Contents of Brown Trout, Rainbow Trout and Perch from Loch Leven, Fife, 1995. Report to Scottish Natural Heritage, Freshwater Fisheries Laboratory, Pitlochry.
Lund, J., 1949. Studies on *Asterionella* 1. The origin and nature of the cells producing seasonal maxima. Journal of Ecology 37: 389–419.
Maitland, P. S., 1979. The distribution of zoobenthos and sediments in Loch Leven, Kinross, Scotland. Archiv für Hydrobiologie 85: 98–125.
Maitland, P. S. & P. M. G. Hudspith, 1974. The zoobenthos of Loch Leven, Kinross and estimates of its production in the sand littoral area during 1970 and 1971. Proceedings of the Royal Society of Edinburgh, B 74: 219–239.
Maitland, P. S., N. W. Charles, N. C. Morgan, K. East & M. C. Gray, 1972. Preliminary research on the production of Chironomidae in Loch Leven, Scotland. In Proceedings of the IBP/UNESCO Symposium on Productivity Problems of Freshwaters. Kazimerz, Dolny, Poland: 795–812.
May, L., 1980a. Ecology of planktonic rotifers at Loch Leven, Kinross-shire. PhD Thesis, Paisley College of Technology.
May, L., 1980b. On the ecology of *Notholca squamula* Müller in Loch Leven, Kinross, Scotland. Hydrobiologia 73: 177–180.
May, L., 1983. Rotifer occurrence in relation to water temperature in Loch Leven, Scotland. Hydrobiologia 104: 311–315.
May, L., 1985. Rotifer sampling—a complete species list from one visit? Hydrobiologia 134: 117–120.
May, L., 1987a. Culturing freshwater, planktonic rotifers on *Rhodomonas minuta* var. *nannoplanctica* Skuja and *Stichococcus bacillaris* Nägeli. Journal of Plankton Research 9: 1217–1223.
May, L., 1987b. Effect of incubation temperature on the hatching of rotifer resting eggs collected from sediments. Hydrobiologia 147: 335–338.
May, L., 1989. Epizoic and Parasitic Rotifers. Hydrobiologia 186(187): 59–67.
May, L., A. E. Bailey-Watts & A. Kirika, 1992. The ecology of *Synchaeta kitina* Rousselet in Loch Leven, Scotland. Hydrobiologia 255(256): 305–315.
May, L., A. E. Bailey-Watts & A. Kirika, 2001. The relationship between *Trichocerca pusilla* (Jennings), *Aulacoseira* spp. and water temperature in Loch Leven, Scotland, UK. Hydrobiologia 446(447): 29–34.
May, L. & L. Carvalho, 2010. Maximum growing depth of macrophytes in Loch Leven, Scotland, United Kingdom, in relation to historical changes in estimated phosphorus loading. Hydrobiologia 646: 123–131.
May, L. L. H. Defew, H. Bennion & A. Kirika, 2011. Historical changes (1905–2005) in external phosphorus loads to Loch Leven, Scotland, UK. Hydrobiologia. doi:10.1007/s10750-011-0922-y.
May, L., I. D. M. Gunn & A. E. Bailey-Watts, 1993. Zooplankton of Loch Leven, Kinross-shire, Scotland. Report to Scottish Natural Heritage.
May, L. & D. H. Jones, 1989. Does interference competition from *Daphnia* affect *Keratella cochlearis* populations in Loch Leven? Journal of Plankton Research 11: 445–461.
May, L. & B. M. Spears, 2011. A history of scientific research at Loch Leven, Kinross, Scotland. Hydrobiologia. doi: 10.1007/s10750-011-0929-4.
Morgan, N. C., 1970. Changes in the fauna and flora of a nutrient enriched lake. Hydrobiologia 35: 545–553.
Morgan, N. C., 1974. Historical background to the International Biological Programme project at Loch Leven, Kinross. Proceedings of the Royal Society of Edinburgh (B) 74: 45–55.
Morrison, B., 1997. Analysis of the Stomach contents of Brown Trout, Rainbow Trout and Perch from Loch Leven, 1996. Report to Scottish Natural Heritage, Freshwater Fisheries Laboratory, Pitlochry.
Moss, B., D. Stephen, C. Alvarez, E. Becares, W. van de Bund, S. E. Collings, E. van Donk, E. de Eyto, T. Feldman, C. Fernández-Alvarez, M. Fernández-Alvarez, R. J. M. Franken, F. Garcia-Criado, E. M. Gross, M. Gyllström, L.-A. Hanssson, K. Irvine, A. Järvalt, J.-P. Jensen, E. Jeppesen, T. Kairesalo, R. Kornijów, T. Krause, H. Künnap, A. Laas, E. Lill, B. Lorens, H. Luup, M. R. Miracle, P. Nõges, T. Nõges, M. Nykänen, I. Ott, W. Peczula, E. T. H. M. Peeters, G. Phillips, S. Romo, V. Russell, J. Salujõe, M. Scheffer, K. Siewertsen, H. Smal, C. Tesch, H. Timm, L. Tuvikene, I. Tonno, T. Virro, E. Vicente & D. Wilson, 2003. The Determination of Ecological Status in Shallow Lakes—a Tested System (ECOFRAME) for Implementation of the European Water Framework Directive. Aquatic Conservation 13: 507–552.
O'Hare, M. T., A. Tree, M. W. Neale, K. Irvine, I. D. M. Gunn, R. Jones & T. Clarke, 2007. Lake benthic macroinvertebrates I: improving sampling methodology. Environment Agency Science Report SC030294/SR1. Environment Agency, Bristol.
Potter, D. W. B. & M. A. Learner, 1974. A study of the benthic macro-invertebrates of a shallow eutrophic reservoir in South Wales with emphasis on the Chironomidae (Diptera); their life-histories and production. Archiv für Hydrobiologie 74: 186–226.
Rosenberg, D. M. & V. H. Resh, 1993. Freshwater Biomonitoring and Benthic Macroinvertebrates. Chapman & Hall, London.
Royal Society of Edinburgh, 1974. The Loch Leven IBP Project. A symposium sponsored by the Royal Society of Edinburgh at the University of Stirling on 11–13 June 1973. Proceedings of the Royal Society of Edinburgh, B 74: 45–421.
Ruttner-Kolisko, A., 1974. Plankton Rotifers: Biology and Taxonomy. Die Binnengĕwasser 26, supplement. Gebrüder Ranz, Dietenkeim: 146 pp.
Saether, O. A., 1979. Chironomid communities as water quality indicators. Holoarctic Ecology 2: 65–74.
Solimini, A. G., G. Free, I. Donohue, K. Irvine, M. Pusch, B. Rossaro, L. Sandin & A. C. Cardoso, 2006. Using Benthic

Macroinvertebrates to Assess Ecological Status of Lakes Current Knowledge and Way Forward to Support WFD Implementation. Office for Official Publications of the European Communities, Luxembourg.

Spears, B. M., L. Carvalho, R. Perkins, A. Kirika & D. M. Paterson, 2011. Long-term variation and regulation of internal phosphorus loading in Loch Leven. Hydrobiologia. doi:10.1007/s10750-011-0921-z.

Sykes, J. M., A. M. J. Lane & D. G. George, 1999. The United Kingdom Environmental Change Network: Protocols for Standard Measurements at Freshwater Sites. Centre for Ecology and Hydrology.

White, J. & K. Irvine, 2003. Littoral meso-habitats in lakes: differences in their macroinvertebrate assemblages and their potential inclusion in assessing ecological quality. Aquatic Conservation: Marine and Freshwater Systems 13: 331–351.

Wiederholm, T., 1981. Associations of lake-living Chironomidae—a cluster analysis of Brundins and recent data from Swedish lakes. Schweizerische Zeitschrift für Hydrologie 43: 140–150.

Williams, D. D. & B. W. Feltmate, 1992. Aquatic Insects. CAB International, Wallingford.

Wilson, R. S. & L. Ruse, 2005. A guide to the identification of genera of Chironomidae pupal exuviae occurring in Britain and Ireland (including common genera from northern Europe) and their use in monitoring of lotic and lentic fresh waters. Freshwater Biological Association.

Winfield, I. J., C. E. Adams, J. D. Armstrong, R. Gardiner, A. Kirika, J. Montgomery, B. M. Spears, D. C. Stewart, J. E. Thorpe & W. Wilson, 2011. Changes in the fish community of Loch Leven: untangling anthropogenic pressures. Hydrobiologia. doi:10.1007/s10750-011-0925-8.

Yang, Y., L. May, I. Gunn, X. Huang & J. Liu. 1999. Comparative Studies on Effects of Predation by Fish on Daphnia in Lake Donghu (China) and Loch Leven (Scotland). Ecosystem Approaches for Fisheries Management, Alaska Sea Grant College Program, AK-SG-99-01: 265–281.

Hydrobiologia (2012) 681:73–84
DOI 10.1007/s10750-011-0925-8

LOCH LEVEN RESEARCH

# Changes in the fish community of Loch Leven: untangling anthropogenic pressures

**I. J. Winfield · C. E. Adams · J. D. Armstrong · R. Gardiner · A. Kirika · J. Montgomery · B. M. Spears · D. C. Stewart · J. E. Thorpe · W. Wilson**

Published online: 15 November 2011


**Abstract** Loch Leven, U.K., contains brown trout (*Salmo trutta*), eel (*Anguilla anguilla*), minnow (*Phoxinus phoxinus*), perch (*Perca fluviatilis*), pike (*Esox lucius*) and three-spined stickleback (*Gasterosteus aculeatus*), with brook lamprey (*Lampetra planeri*) and stone loach (*Barbatula barbatula*) also present in its tributaries. Arctic charr (*Salvelinus alpinus*), Atlantic salmon (*Salmo salar*) and flounder (*Platichthys flesus*) are now extinct. The brown trout population has supported a world-renowned recreational fishery for over a century, although a decline in fishery performance led to extensive stocking between 1983 and 2006, including with non-native rainbow trout (*Oncorhynchus mykiss*). This review combines historical information with contemporary gill-net and hydroacoustic surveys. In 2008, brown trout, perch and three-spined sticklebacks were abundant, but pike and stone loach were rare. The obstruction of migratory routes was probably responsible for the loss of Atlantic salmon and flounder, while a lowering of water level likely caused the extinction of Arctic charr and contributed to a reduction in pike abundance. Perch abundance has fluctuated markedly, being influenced by disease and eutrophication, although a reduction in nutrients and associated recovery of macrophytes are likely to have benefitted this species. Although the brown trout population has undoubtedly shown a long-term decline, individuals are currently in excellent condition.

**Keywords** Population · Eutrophication · Lake level · Fishery · Brown trout (*Salmo trutta*) · Rainbow trout (*Oncorhynchus mykiss*) · Perch (*Perca fluviatilis*) · Cormorant (*Phalacrocorax carbo*)

Guest editors: L. May & B. M. Spears / Loch Leven: 40 years of scientific research

I. J. Winfield (✉)
Centre for Ecology & Hydrology, Lancaster Environment Centre, Library Avenue, Bailrigg, Lancaster LA1 4AP, UK
e-mail: ijw@ceh.ac.uk

C. E. Adams
Scottish Centre for Ecology and the Natural Environment, University of Glasgow, Glasgow G12 8QQ, UK

J. D. Armstrong · R. Gardiner · D. C. Stewart · J. E. Thorpe
Freshwater Laboratory, Marine Scotland, Faskally, Pitlochry PH16 5LB, UK

A. Kirika · B. M. Spears
Centre for Ecology & Hydrology, Bush Estate, Penicuik, Midlothian EH26 0QB, UK

J. Montgomery · W. Wilson
Kinross Estate Office, Kinross KY13 8AS, UK

## Introduction

Along with phytoplankton, which may periodically appear to colour an entire lake vivid green, and birds, which at certain times of year teem on its water surface, terrestrial margins and in the skies above it, a lake's fish populations commonly enjoy the highest public profiles of all of its flora and fauna. However, in contrast to phytoplankton, which are easy to sample, and birds, which are easy to count, the assessment and monitoring of lake fish communities is technically difficult and presents a number of continuing methodological challenges (Kubečka et al., 2009). When the complexities and uncertainties of fisheries socio-economics within a multi-use landscape (e.g. Arlinghaus et al., 2002) are added to this situation, objective and confident understanding and management of lake fish communities become extremely difficult to achieve.

The history and contemporary state of the fish community of Loch Leven in east-central Scotland, U.K., is a good example of this complexity, even though a total of only 12 resident and migratory fish species has been recorded in this large, shallow and eutrophic lake (see below). Amongst these species, the local population of brown trout (*Salmo trutta*) has received by far the greatest attention in terms of research effort and fisheries exploitation, with commercial netting having been practised here from 1314 to 1873 and recreational fly fishing from 1844 onwards (Thorpe, 1974a; Munro, 1994). The latter, at its height, supported 52 hire boats (W. W., unpublished data). Management activities in support of angling activity have been documented by Montgomery (1994), including an early example of the use of hatchery and rearing ponds in 1882 when 60,000 fry and 4,000 2-year-old brown trout of unrecorded origin were placed in inflowing streams, a practice that was expanded from the 1920s to 1936 resulting in up to 300,000 fry being hatched each year. Subsequently, the hatchery was closed during the Second World War and only reopened in the early 1980s using 'home-bred' stock of exclusively local origin (Montgomery, 1994). The Loch Leven trout fishery persists to the present with 13 boats available for hire (W. W., unpublished data) and a national and international reputation such that it was included as a venue for the 29th FIPS-Mouche World Fly Fishing Championships in June 2009. In recent years, however, it has experienced turbulent fortunes with various anthropogenic pressures (e.g. Duncan, 1994), the socio-economic vagaries of the angling community (e.g. Montgomery, 1994) and predation by cormorants (*Phalacrocorax carbo*) (e.g. Stewart et al., 2005) all potentially involved. Many of these influential factors have a long local history (May & Spears, 2011) and so an understanding of the contemporary fish community requires a correspondingly long-term perspective.

This article reviews historical published and unpublished information in combination with contemporary gill net and hydroacoustic surveys to describe long-term changes in the fish community of Loch Leven. Observed alterations in the fish community are then interpreted in the context of historical and current anthropogenic pressures.

## Materials and methods

### Study site

Loch Leven is a large (surface area approximately 13.3 km$^2$) but relatively shallow (mean depth 3.9 m, maximum depth 25.5 m) lake located in east-central Scotland, U.K. (56° 12′N, 3° 22′W; altitude 107 m). This culturally eutrophicated water body drains a catchment of 145 km$^2$ and discharges into the Forth Estuary via the River Leven. Further details of its bathymetry (Kirby, 1971), water quality history (Carvalho et al., 2011; May et al., 2011; Spears et al., 2011), phytoplankton (Bailey-Watts, 1982; Bailey-Watts et al., 1990), zooplankton (Gunn et al., 2011), macrophytes (Dudley et al., 2011), macroinvertrbrates (Gunn et al., 2011) and birds (Carss et al., 2011) are reported elsewhere.

### Historical published and unpublished information

Published and unpublished information relevant to an understanding of the fish community of Loch Leven was collated by searches of the published and grey literature, combined with unpublished data held by the authors who collectively have been directly responsible for, or involved with, almost all fish research and all fisheries operations on the loch. This resulted in the identification of over 30 publications and reports including, or supplemented with, data on fish populations from the 1960s, 1970s, 1980s, 1990s and 2000s and fisheries data from 1900 to the 2000s.

### Contemporary hydroacoustic and gill net surveys

Before this study, the most recent sampling of the fish populations of Loch Leven had been undertaken using survey gill nets during March and December 2001 (Stewart et al., 2005). Consequently, a community sampling programme using a combination of hydroacoustics and survey gill nets as used on a series of other lochs by Winfield et al. (2009a) was begun in late 2007.

On 10 October 2007, day- and night-time hydroacoustic surveys were performed where water depth exceeded approximately 5 m in the southern area of the loch using a BioSonics DT6000 echo sounder (BioSonics Inc., Seattle, U.S.A.) with a 200 kHz split-beam vertical transducer following the approach described by Winfield et al. (2009a). The transducer was positioned approximately 0.5 m below the surface of the water and data were recorded starting from a range of 2 m. Following a standard hydroacoustic analysis employed extensively elsewhere in the U.K. by Winfield et al. (in press), the resulting data were used to produce densities of 'small' (i.e. −52 to −45 dB, length 40–99 mm), 'medium' (−44 to −37 dB, length 100 249 mm) and 'large' (greater than −37 dB, length greater than 250 mm) fish along each of 18 transects.

Gill netting was undertaken using benthic and pelagic versions of the Norden survey gill net as described by Appelberg (2000). The benthic version is bottom-set and is approximately 1.5 m deep and 30 m long, with 12 panels of equal length having bar-mesh sizes 5, 6.25, 8, 10, 12.5, 15.5, 19.5, 24, 29, 35, 43 and 55 mm, respectively. The pelagic version, which is set floating from the water surface, is approximately 6.0 m deep and 27.5 m long, with 11 panels of equal length having bar-mesh sizes 6.25, 8, 10, 12.5, 15.5, 19.5, 24, 29, 35, 43 and 55 mm, respectively. A total of nine such nets was set overnight to sample three inshore-bottom sites, three offshore-bottom sites and three offshore-surface sites on 17 March, 20 August and 6 October 2008. Upon lifting, all fish were removed from the nets and killed before being identified, counted and then frozen at −20°C for future processing in the laboratory. The latter comprised measurements of individual length (fork length, mm) and weight (g) and the removal of materials for age and diet determination.

For comparison purposes, corresponding individual data were assembled from brown trout and perch taken from a total of seven other lochs sampled using the same gill-netting methodology in 2007 and 2008 by Winfield et al. (2008a, b, 2009b), and from Loch Leven perch sampled in March 2001 (D. C. S., unpublished data from gill netting) and August 1968 (J. E. T., unpublished data from trawling).

## Results

### Historical fish community

A total of 11 native fish species has been recorded in Loch Leven, i.e. Arctic charr (*Salvelinus alpinus*), Atlantic salmon (*Salmo salar*), brook lamprey (*Lampetra planeri*), brown trout, eel (*Anguilla anguilla*), flounder (*Platichthys flesus*), minnow (*Phoxinus phoxinus*), perch (*Perca fluviatilis*), pike (*Esox lucius*), stone loach (*Barbatula barbatula*) and three-spined stickleback (*Gasterosteus aculeatus*). However, brook lamprey and stone loach have always been largely restricted to tributary streams and flounder was probably only an occasional visitor (Day, 1887).

Three of the above species became locally extinct during historical times. The Atlantic salmon and flounder, which both used to migrate along the outflowing River Leven to and from the Forth Estuary, were lost during the eighteenth century (Day, 1887), while Arctic charr have not been recorded since 1837 (Burns-Begg, 1874).

Only one fish species has subsequently been added to the Loch Leven fish community, i.e. the non-native rainbow trout (*Oncorhynchus mykiss*), which was stocked for fishery purposes from 1993 to 2004.

### Brown and rainbow trout fishery

The performance of the Loch Leven brown trout rod fishery from 1900 to 2002 is given in Fig. 1a; after 2002, catch records collected by the fishery became less reliable as an indicator of fishery performance as a result of anglers releasing fish that they could have retained. Between 1900 and 1970, the annual catch frequently exceeded 40,000 brown trout with less than 20,000 fish landed in only a very few years. After 1970, however, catches declined markedly and rarely exceeded 20,000 fish. Furthermore, this level of brown trout catch was last surpassed in 1990. The post-1970 decline in absolute catch of brown trout was

**Fig. 1** Loch Leven fishery **a** absolute catch for brown trout (*closed circles, continuous line*) and rainbow trout (*open circles, broken line*) from 1900 to 2002, **b** average weight for brown trout (*closed circles, continuous line*) and rainbow trout (*open circles, broken line*) from 1900 to 2002 and **c** catch-per-unit-effort (CPUE) for brown trout (*closed circles, continuous line*) and rainbow trout (*open circles, broken line*) from 1975 to 2002

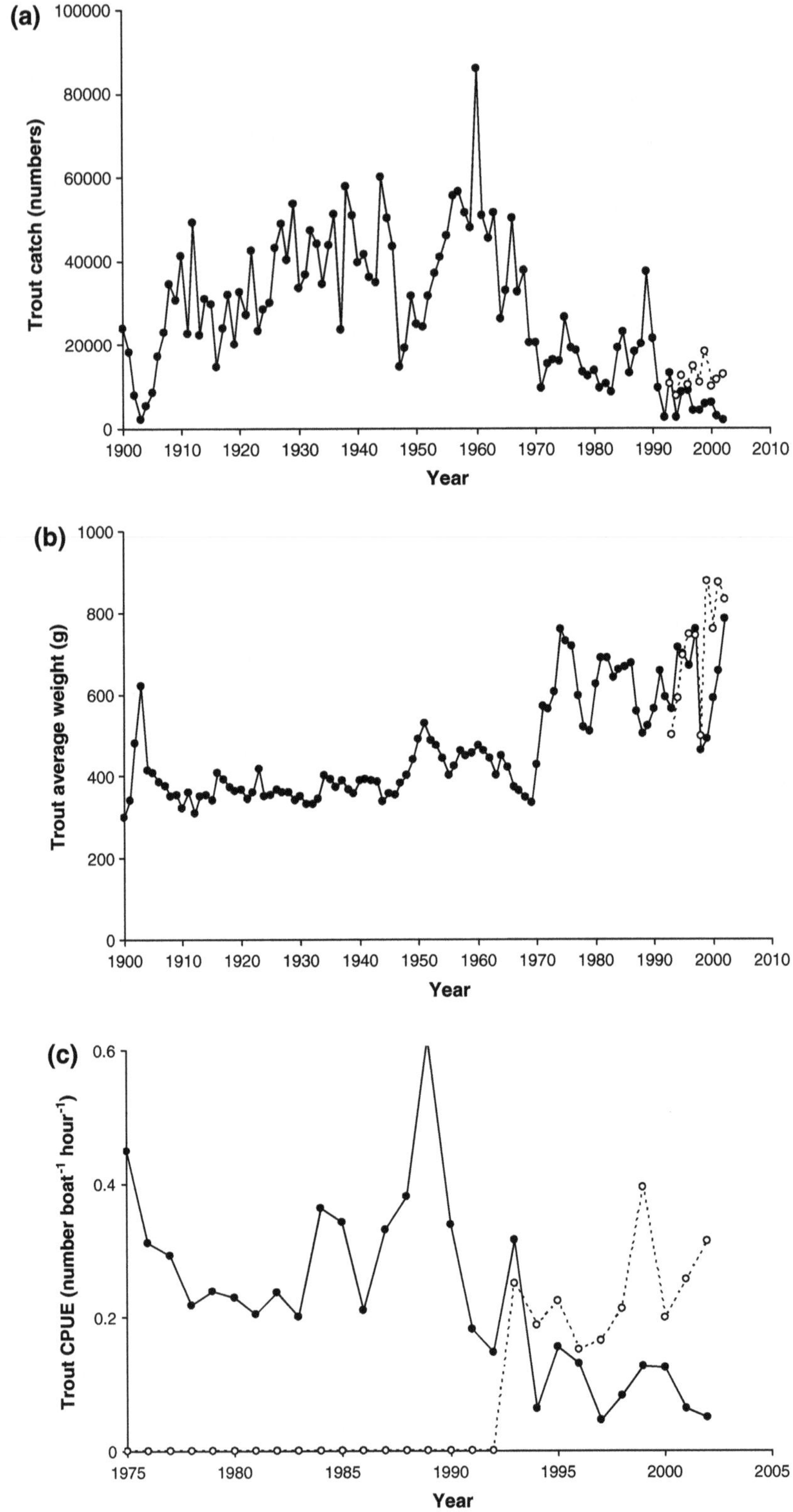

accompanied by a marked increase in individual weight (Fig. 1b).

This dramatic decline in fishery performance led to the reintroduction, in 1983, of brown trout stocking directly into the loch using 5,000 fish of local origin from the re-opened hatchery and rearing-ponds (Montgomery, 1994). The number of brown trout stocked each year steadily increased until 1988, when 166,000 fish were introduced. Stocking in excess of 100,000 brown trout continued in most years until 2004, when the number was reduced to 5,000 and stocking was discontinued in 2006. Unlike the earlier period when the stocked fish consisted mainly of fry introduced to tributary streams, from 1983 onwards the stocked fish were released directly into the loch, generally as yearlings of 120–180 mm in length (Montgomery, 1994). However, from 2004 to 2006, the emphasis was on introducing older fish of approximately 280 mm in length (W.W., unpublished data) which were already large enough to be taken by anglers.

During the 1980s component of the stocking programme, the performance of the brown trout fishery in terms of catch-per-unit-effort (CPUE, calculated from fishery returns from all anglers as the number of brown trout caught per boat-hour) was relatively stable but then decreased again in the 1990s (Fig. 1c). This further deterioration in brown trout fishery performance, coupled with poor environmental conditions (in terms of low oxygen availability and low water clarity) and changing angler preferences, led the fishery to stock with rainbow trout from 1993 to 2004. During this period at least 30,000 fish were stocked each year. This stocking dramatically changed the nature of the Loch Leven fishery such that the absolute catch (Fig. 1a) and CPUE of rainbow trout (calculated as for brown trout, Fig. 1c) almost equalled that of brown trout in the first year of stocking of the latter species, and then greatly exceeded it in all subsequent years with stocking and for which catch data are available.

Perch population

Sampling in 2008 produced a total of 641 perch, of which 45 individuals were taken in March when they ranged between 83 and 251 mm in length and 6 and 292 g in weight. The length frequency distribution of these fish contained only one individual above 180 mm in length (Fig. 2a), which amounted to 2% of individuals and contrasted with corresponding data from March 2001 (Fig. 2b) and August 1968 (Fig. 2c) when 43% of 67 individuals and 48% of 180 individuals, respectively, exceeded this length. This currently truncated length frequency distribution was also evident when the 2008 Loch Leven length and weight data were compared with corresponding data from perch populations in two other lochs sampled using identical methodology although, on the basis of their relatively high weights at lengths, individual fish were in good condition (Fig. 2d).

Brown trout population

Sampling in 2008 produced a total of 163 brown trout, of which 25 individuals were taken in March when they ranged between 135 and 577 mm in length and 27 and 2,317 g in weight. The length frequency distribution of the latter individuals contained relatively high numbers of large fish (Fig. 3a), with a length-weight relationship indicative of excellent individual condition and with markedly higher weights-at-lengths for individuals greater than 400 mm when compared with similar data from several other lochs (Fig. 3b).

Contemporary fish community

The population density of all fish recorded by hydroacoustics in the deeper (i.e. water depth 5 m and greater) areas of Loch Leven during October 2007 was 72.7 fish ha$^{-1}$ (geometric mean with lower and upper 95% confidence limits of 21.8 and 242.5 fish ha$^{-1}$, respectively). Note that these results were subject to a near-surface blind zone of approximately 2.5 m within which fish could not be detected. A breakdown by fish length classes is given in Fig. 4a.

Gill-net sampling in 2008 produced a total of 1,139 individuals of five fish species, i.e. brown trout, perch, pike, three-spined stickleback and stone loach (Fig. 4b). However, catches were numerically dominated by brown trout (14% by numbers), perch (56%) and three-spined sticklebacks (29%), with pike (less than 1%) and stone loach (less than 1%) represented by only single individuals.

## Discussion

The fish community of Loch Leven is relatively simple in terms of species number, with a cumulative total of

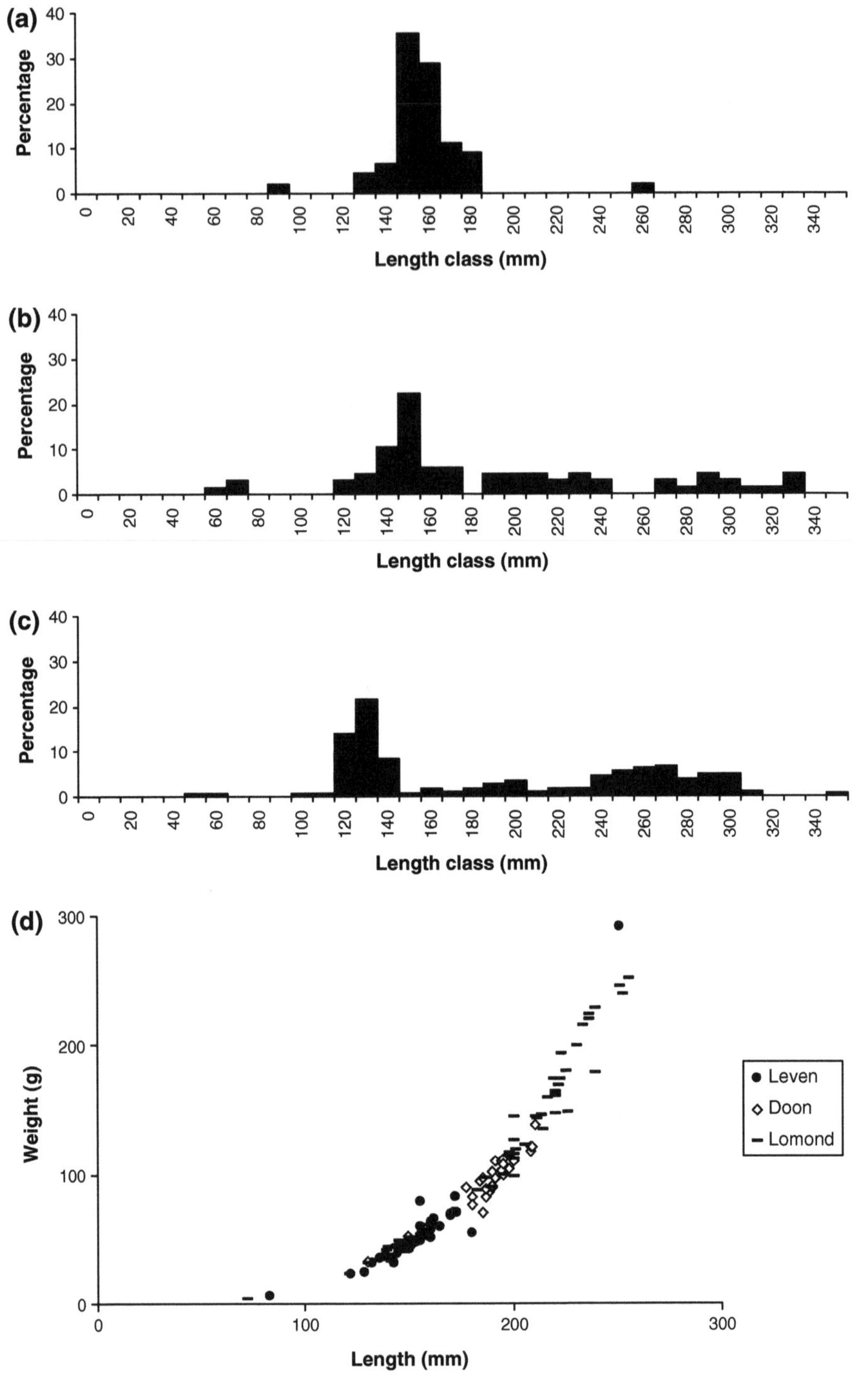


**Fig. 2** Length frequency distributions of perch sampled from Loch Leven **a** 45 perch in March 2008, **b** 67 perch in March 2001, **c** 180 perch in August 1968 and **d** length-weight relationship of 45 perch sampled from Loch Leven in March 2008. For comparison, the latter also shows corresponding data from Loch Doon and Loch Lomond

only 12 species including the rainbow trout which is non-native to the U.K. and was introduced for fishery purposes. However, it has displayed a number of major changes over the long-term and the extensive 2008 sampling recorded only five species, i.e. brown trout, perch, pike, three-spined stickleback and stone loach. Furthermore, in numerical terms, the community is now dominated by just three species, i.e. brown trout, perch and three-spined stickleback, with a total abundance of 72.7 fish $ha^{-1}$ as measured by the initial

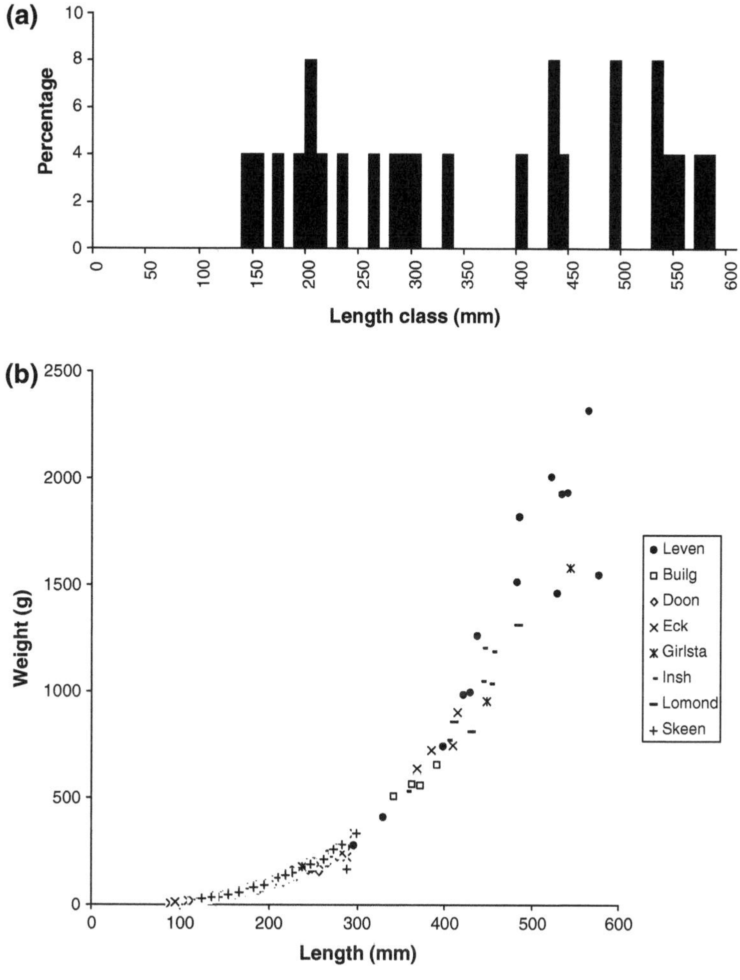


**Fig. 3** **a** Length frequency distribution and **b** length-weight relationship of 25 brown trout sampled from Loch Leven in March 2008. For comparison, the latter also shows corresponding data from Lochs Builg, Doon, Eck, Insh, Lomond and Skeen, and Loch of Girlsta

extensive hydroacoustic survey of October 2007 being near the middle of the range observed for 17 other U.K. water bodies by Winfield et al. (in press).

The present absence of brook lamprey and rarity of stone loach probably reflect the local distribution patterns of these two riverine species, with their populations primarily inhabiting the loch's tributary streams, and so are not of any significance in terms of long-term changes. However, the current apparent absence of Arctic charr, Atlantic salmon, eel (although this species is poorly sampled by gill nets), flounder, minnow and rainbow trout and the great scarcity of pike are worthy of further discussion where data allow, as are the observed changes in the perch and brown trout populations. Unfortunately, most previous assessments of the loch's fish including Thorpe (1974a, b), O'Grady et al. (1993) and Alexander et al. (1999) have not attempted to sample the minnow and three-spined stickleback populations. Consequently, while our 2008 sampling has shown that these two species are, respectively, now apparently absent and relatively abundant, nothing can be concluded concerning their long-term population dynamics.

The present Loch Leven fish community composition is also noteworthy in the context of absent species. First, it appears that most or all of the 370,000 rainbow trout stocked into the loch between 1993 and 2004 have now either died or been removed and that no local reproduction has occurred. Although an extremely small number of juvenile and adult rainbow trout have recently been caught by anglers, these are thought to have entered the system from one or more of several newly opened fisheries on the loch's tributaries (B. M. S. and W. W., personal observations). Second, a number of other major lakes in the

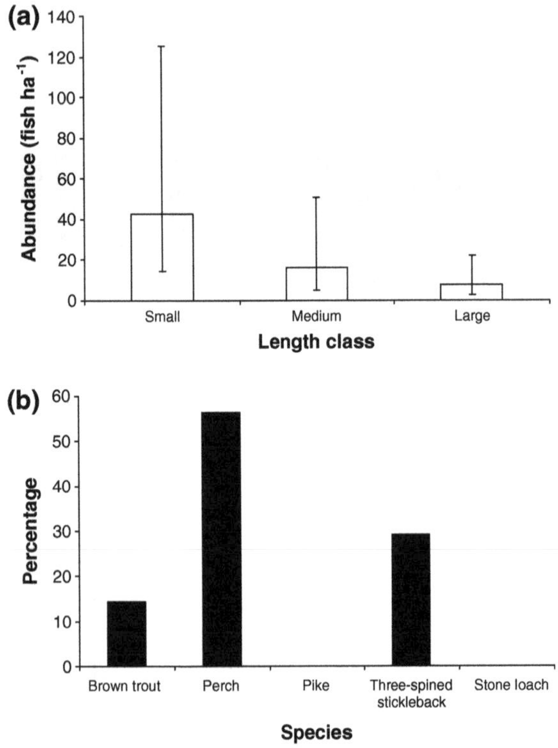


**Fig. 4** **a** Abundance estimates (geometric means with 95% confidence limits) by length class for small (length 40–99 mm), medium (length 100–249 mm) and large (length equal to or greater than 250 mm) fish recorded at Loch Leven on 10 October 2007, **b** Composition by numbers of the contemporary fish community of Loch Leven based on sampling in March, August and October 2008. Total sample size was 1,139 fish. Note that single pike and stone loach were sampled, although they are not visible on the figure at this scale

U.K. have recently experienced unconsented introductions of several fish species, most notably roach (*Rutilus rutilus*) and ruffe (*Gymnocephalus cernuus*), as a result of anglers fishing for pike using fish as live baits (Winfield & Durie, 2004). Although there is thus a potential for such unwanted introductions to Loch Leven, with potential implications for the native fish community, the extensive sampling of 2008 did not record any new arrivals.

The three apparent local extinctions of fish species, i.e. Arctic charr, Atlantic salmon and flounder, all occurred before scientific studies of Loch Leven began, so their causes can only be speculated upon. However, Burns-Begg (1874) noted that the Arctic charr was lost only a few years after the marked lowering of the loch level in 1830, which is strong circumstantial evidence that this was the major factor behind its extinction. This potential pressure has parallels with more recent and better-studied effects of falls in lake levels on Arctic charr and other members of the fish community of Haweswater, U.K. (Winfield et al., 1998), and elsewhere (Maitland et al., 2007). Furthermore, even if the Arctic charr had survived this pressure, it would then have faced a significant pressure from subsequent eutrophication which currently stresses other U.K. populations such as those in Windermere (Winfield et al., 2008c). The local extinctions of Atlantic salmon and flounder occurred even earlier in the eighteenth century and may have been due to the obstruction of migratory routes by river pollution, although increasing physical difficulty of passage due to sluice construction associated with the subsequent lowering of the loch in 1830 and the construction of weirs further downstream would probably have formed additional factors.

The above local anthropogenic pressures together with more widespread factors (Bonhommeau et al., 2008) are also likely to have contributed to a decline in the eel population of Loch Leven. The catches of commercial eel fishings at the outflow from the loch declined through the nineteenth century and these operations finally came to an end in 1901 (Munro, 1994). Furthermore, this species was not recorded during extensive sampling from 1968 to 1973 (J. E. T., personal observation). Even allowing for the inefficient sampling of eels by gill nets, it is remarkable that the survey of O'Grady et al. (1993) and the present sampling also failed to record any individuals of this once ubiquitous species. However, recent very infrequent captures by anglers, the sighting of a dead eel in the harbour at Kinross and very occasional captures of eels in survey work in the Gairney Burn, a tributary of Loch Leven, (B. M. S., W. W. and R. G., personal observations) confirm the presence of at least a few eels. It is possible that the eels now present may have accessed the catchment via the River Devon, which runs independently into the Firth of Forth, and is linked with the Gairney Burn by the Pow Burn which drains into both catchments.

The pike population of the loch has also declined markedly since historic times, with the surveys of O'Grady et al. (1993) and this study recording 0 and 1 individuals, respectively. Duncan (1994) considered that a long-term pike removal programme operated by the fishery was probably responsible for this decline, although a reduction in macrophytes associated with eutrophication over the same period (Dudley et al., 2011), which are used as spawning and nursery habitat by pike (Craig, 1996), may also have been involved. In addition, the earlier lowering of water levels may also have had a negative effect on pike recruitment through reducing spawning and nursery habitat.

The reduction in macrophyte abundance mentioned above is also likely to have had a negative effect on the perch population of Loch Leven because this species uses their physical complexity for spawning and nursery habitat and it also influences competitive interactions involving adult members of the population (Craig, 2000). In addition, Duncan (1994) considered that a perch disease that he reported as first observed in Loch Leven in 1983 may also have contributed to a marked decline, although the first signs of this disease were actually observed over a decade earlier in 1971 (J. E. T., personal observation). Long-standing and marked fluctuations in the perch population are evident from abundance estimates of 968,000 individuals and 253,000 individuals in 1970 and 1971, respectively (Thorpe, 1974a), by the scarcity of perch in the 1993 sampling of O'Grady et al. (1993), and by their limited size distribution in the present sampling of 2008. Even allowing for the substantial natural variation in abundance of perch observed in relatively undisturbed habitats such as that of the north basin of Windermere (Paxton et al., 2004), the decline of the perch population in the eutrophicated Loch Leven has been particularly marked. However, some recovery of the perch population might now be expected as the loch's level of eutrophication reduces (Carvalho et al., 2011) and its macrophytes increase in abundance (Dudley et al., 2011; May & Carvalho, 2010). As adult perch consume over 6% of their body weight per day during the summer (Thorpe, 1977) and young perch are probably a major planktivore within the Loch Leven system, such population changes are likely to have important implications for the loch's food web through impacts on their prey populations.

The brown trout population has undoubtedly received by far the most research attention of all of Loch Leven's fish species, driven in large part by its fishery importance and thus high public profile. Although documented and undocumented changes over time in fishing practices and equipment mean that trends in fishery performance must be interpreted with caution, the decline of the Loch Leven trout fishery has been so marked that it undoubtedly reflects an actual decline in the local brown trout population. This problem was widely acknowledged by the early 1990s and led to much previous research being presented or reviewed during a dedicated conference on the trout fishery reported by Hutchinson & Walker (1994). Consequently, this earlier research will not be revisited here in any detail.

Brown trout spawn in flowing water and the streams that feed the loch are therefore a vital component of the habitat for this species. Despite heavy impacts of agriculture and some localised gravel abstraction on parts of the catchment, the streams entering Loch Leven are highly productive of brown trout. Surveys conducted in the early 1990s (Armstrong et al., 1994; Armstrong & Gardiner, 1995) assessed the strength and aspects of the demographics of trout in the streams. Strong relationships were evident with key components of the habitat such that the degree of bank overhangs (which correlated strongly with bankside vegetation) together with coarse in-stream substratum explained 76% of the variation in trout densities in winter. Trout densities in autumn were strongly influenced by the degree of overhead cover together with densities of trout in May/June. Trout older than 1+ in autumn were scarce in the sampled sites. These results were consistent with a possible limitation of spawning on production, through one or a combination of low numbers of eggs, or insufficient distribution of spawning substratum. There was no evidence of trout tending to remain in the streams as older fish rather than migrating to the loch, as observed also in an earlier study (Thorpe, 1974a). Despite this potential limitation of recruitment on production, densities of trout were high in Loch Leven streams compared with other populations and at least as high as those determined by Thorpe (1974a). The presence of the remnants of yolk sacs and a wide range of lengths in samples collected in May/June were consistent with the occurrence of a wide range of spawning times and may facilitate high levels of production through heterogeneous advantage (Griffiths & Armstrong, 2001) due to young fish of different sizes using different local niches and therefore competing little with one another. Further examination of the streams would be enhanced by trapping of trout migrating to and from the loch to measure production of smolts and migrant fry directly, to assess the numbers of spawners, and to derive indices of stock size.

Estimates of the brown trout population within Loch Leven have been undertaken during three different periods, including an examination of the effect of an aeromonad epidemic on the spawning population (Thorpe & Roberts, 1972). However, caution is urged with respect to the interpretation of long-term dynamics because outside of the population estimates by Thorpe (1974a), which themselves ranged from 127,000 fish in 1968 to 52,000 fish in 1971, later assessments by O'Grady et al. (1993) and Alexander et al. (1999) were undertaken using different techniques, for different components of the population and for different times of the year. Consequently, their assembly into a single time series cannot be performed robustly.

These latter data have subsequently been used by both Wright (2003) and Stewart et al. (2005) to assess the controversial subject of the impact of cormorants overwintering at Loch Leven. In common with other inland locations across the U.K., the number of cormorants wintering at Loch Leven increased substantially in the late 1980s (Wright, 2003). By comparing CPUE for gill-net and angling catches before and after the arrival of wintering cormorants, Wright (2003) concluded that cormorants had no effect on brown trout abundance or fishery performance. Using diet analysis coupled with a consumption model, Stewart et al. (2005) estimated that cormorants consumed 80,803 brown and 5,213 rainbow trout over a 7 month period and concluded that there was a high potential for competition between the birds and the fishery for available fish. Stewart et al. (2005) also found a link between brown trout stocking level in spring and cormorant abundance the following winter, suggesting that increased stocking could have attracted more cormorants to the site and that the subsequent increase in predation would cancel out any benefits of stocking on the fishery yield.

The numbers of brown and rainbow trout estimated to have been removed by cormorants (Stewart et al., 2005) suggested high levels of impact relative to any of the estimates of the trout populations that have been attempted (Wright, 2003). However, even in Loch Leven, which has been more intensively studied than most other large water bodies in the U.K., there is considerable uncertainty regarding the size of the trout populations such that precise estimates of the impacts of piscivorous birds on them are not possible.

A clear feature of the brown trout population is that the length frequency distribution and length-weight relationship observed in 2008 indicates that those individuals present are in excellent individual condition. This observation is consistent with the marked increase in the average weight of brown trout in the fishery, which accompanied the decline in their apparent abundance in the 1970s. This relationship is consistent with density-dependent increase in size rather than reduction in density due to mortality associated with poor average condition.

## Conclusion

The fish community of Loch Leven is relatively simple with a cumulative total of 12 species, but it has undergone a number of extinctions and major changes over historical and recent times and is currently dominated by just brown trout, perch and three-spined stickleback. A lowering of water level probably caused the local extinction of Arctic charr and contributed to the deterioration of the pike population, while the obstruction of migratory routes was likely responsible for the loss of Atlantic salmon and flounder. The perch population has shown marked fluctuations in abundance typical of this species, although its local dynamics have also probably been influenced by a perch-specific disease outbreak and increasing eutrophication. The present lowering of nutrient levels in the loch and associated recovery of macrophytes are likely to benefit this species, which, given its position as a major planktivore within the system, is likely to have important implications for the loch's food web. The size of the brown trout population was initially estimated by application of mark and recapture methodology on an unusually large scale. However, differences in methodologies applied subsequently make it difficult to compare the size of the population over time and this has hampered robust analysis of the factors that may be involved in the demise of the trout fishery. Nonetheless, the decline has been so marked that it undoubtedly reflects an actual decline in the local brown trout population.

**Acknowledgments** We would like to thank Janice Fletcher, Ben James and other colleagues at the Centre for Ecology & Hydrology for help in the field and laboratory during the recent study of the fish community of Loch Leven. Colin Bean of Scottish Natural Heritage kindly allowed the use of individual brown trout and perch data collected under contract at a number of other locations. We also gratefully acknowledge appropriate sampling permissions from the Kinross Estates, Scottish Natural Heritage and The Scottish Government. The preparation of this review was primarily funded by the Natural Environment Research Council. Loch Leven is part of the UK Environmental Change Network (http://www.ecn.ac.uk/).

## References

Alexander, G., C. E. Adams, J. Devine, J. Drummond & C. W. Bean, 1999. Loch Leven: hydroacoustic technique development and fish population estimation. University of Glasgow Report to Scottish Natural Heritage. Scottish Natural Heritage, Edinburgh.

Appelberg, M., 2000. Swedish standard methods for sampling freshwater fish with multi-mesh gillnets. Fiskeriverket Information 1: 3–32.

Arlinghaus, R., T. Mehner & I. G. Cowx, 2002. Reconciling traditional inland fisheries management in industrialized countries, with emphasis on Europe. Fish and Fisheries 3: 261–316.

Armstrong, J. D. & W. R. Gardiner, 1995. Juvenile trout populations in the Loch Leven catchment. Fisheries Research Services Report 13/95.

Armstrong, J. D., W. R. Gardiner & A. F. Walker, 1994. Juvenile brown trout production in the Loch Leven catchment. Fisheries Research Services Report 07/94.

Bailey-Watts, A. E., 1982. The composition and abundance of phytoplankton in Loch Leven (Scotland) 1977–1979 and a comparison with the succession in earlier years. Internationale Revue der Gesamten Hydrobiologie 67: 1–25.

Bailey-Watts, A. E., A. Kirika, L. May & D. H. Jones, 1990. Changes in phytoplankton over various time scales in a shallow eutrophic lake: the Loch Leven experience with special reference to the influence of flushing rate. Freshwater Biology 23: 85–111.

Bonhommeau, S., E. Chassot, B. Planque, E. Rivot, A. H. Knap & O. Le Pape, 2008. Impact of climate on eel populations of the Northern hemisphere. Marine Ecology Progress Series 373: 71–80.

Burns-Begg, R., 1874. The Loch Leven angler. Barnett, Kinross.

Carvalho, L., C. Ferguson, B. M. Spears, I. D. M.Gunn, H. Bennion, A. Kirika & L. May, 2011. Water quality of Loch Leven: responses to enrichment, restoration and climate change. Hydrobiologia. doi:10.1007/s10750-011-0923-x.

Carss, D., B. M. Spears, L. Quinn & R. Cooper, 2011. Long-term variations in waterfowl populations in Loch Leven: identifying discontinuities between local and national trends. Hydrobiologia. doi:10.1007/s10750-011-0927-6.

Craig, J. F., 1996. Pike: biology and exploitation. Chapman & Hall Fish and Fisheries Series 19. Chapman & Hall, London.

Craig, J. F., 2000. Percid fishes: systematics, ecology and exploitation. Blackwells, Oxford.

Day, F., 1887. On the Loch-Leven trout (*Salmo levenensis*). Journal of the Linnaean Society of London, Zoology 20: 71–90.

Dudley, B., I. D. M. Gunn, L. Carvalho, I. Proctor, M. T. O'Hare, K. J. Murphy, A. Milligan, 2011. Changes in aquatic macrophyte communities in Loch Leven—evidence of recovery from eutrophication? Hydrobiologia. doi:10.1007/s10750-011-0924-9.

Duncan, W., 1994. The fish community in Loch Leven. In Hutchinson, P. & A. Walker (eds), The Loch Leven trout fishery—its future. Institute of Fisheries Management, Pitlochry: 29–38.

Griffiths, S. W. & J. D. Armstrong, 2001. The benefits of genetic diversity outweigh those of kin association in a territorial animal. Proceedings of the Royal Society London, Series B 268: 1293–1296.

Gunn, I. D. M., M. T. O'Hare, P. S. Maitland & L. May, 2011. Long-term trends in Loch Leven invertebrate communities. Hydrobiologia. doi:10.1007/s10750-011-0926-7

Hutchinson, P. & A. Walker (eds), 1994. The Loch Leven trout fishery—its future. Institute of Fisheries Management, Pitlochry.

Kirby, R. P., 1971. The bathymetrical resurvey of Loch Leven, Kinross. Geographical Journal 137: 372–378.

Kubečka, J., U. S. Amarasingh, S. A. Bonar, J. Hateley, P. Hickley, E. Hohausová, J. Matěna, J. Peterka, P. Suuronen, V. Tereschenko, R. Welcomme & I. J. Winfield, 2009. The true picture of a lake or reservoir fish stock: a review of needs and progress. Fisheries Research 96: 1–5.

Maitland, P. S., I. J. Winfield, I. D. McCarthy & F. Igoe, 2007. The status of Arctic charr *Salvelinus alpinus* in Britain and Ireland. Ecology of Freshwater Fish 16: 6–19.

May, L. & L. Carvalho, 2010. Maximum growing depth of macrophytes in Loch Leven, Scotland, United Kingdom, in relation to historical changes in estimated phosphorus loading. Hydrobiologia 646: 123–131.

May, L. & B. M. Spears, 2011. A history of scientific research at Loch Leven, Kinross, Scotland. Hydrobiologia. doi: 10.1007/s10750-011-0929-4

May, L. L. H. Defew, H. Bennion & A. Kirika, 2011. Historical changes (1905–2005) in external phosphorus loads to Loch Leven, Scotland, UK. Hydrobiologia. doi:10.1007/s10750-011-0922-y

Montgomery, D., 1994. Management of the Loch Leven trout fishery. In Hutchinson, P. & A. Walker (eds), The Loch Leven trout fishery—its future. Institute of Fisheries Management, Pitlochry: 26–28.

Munro, D., 1994. Loch Leven and the River Leven: a landscape transformed. The River Leven Trust, Markinch.

O'Grady, M. F., P. Gargan & W. Roche, 1993. A fish stock survey of Loch Leven and management proposals for this resource as a trout fishery. Central Fisheries Board Report to Scottish Natural Heritage. Scottish Natural Heritage, Edinburgh.

Paxton, C. G. M., I. J. Winfield, J. M. Fletcher, D. G. George & D. P. Hewitt, 2004. Biotic and abiotic influences on the recruitment of perch (*Perca fluviatilis*) in Windermere, U. K. Journal of Fish Biology 65: 1622–1642.

Spears B. M., L. Carvalho, R. Perkins, A. Kirika & D. M. Paterson, 2011. Long-term variation and regulation of internal phosphorus loading in Loch Leven. Hydrobiologia. doi: 10.1007/s10750-011-0921-z.

Stewart, D. C., S. J. Middlemas, W. R. Gardiner, S. Mackay & J. D. Armstrong, 2005. Diet and prey selection of cormorants

(*Phalacrocorax carbo*) at Loch Leven, a major stocked trout fishery. Journal of the Zoological Society of London 267: 191–201.

Thorpe, J. E., 1974a. Trout and perch populations at Loch Leven, Kinross. Proceedings of the Royal Society of Edinburgh, B 74: 296–313.

Thorpe, J. E., 1974b. Estimation of the number of brown trout *Salmo trutta* (L.) in Loch Leven, Kinross, Scotland. Journal of Fish Biology 6: 135–152.

Thorpe, J. E., 1977. Daily ration of adult perch, *Perca fluviatilis* L. during summer in Loch Leven, Scotland. Journal of Fish Biology 11: 55–68.

Thorpe, J. E. & R. J. Roberts, 1972. An aeromonad epidemic in the brown trout (*Salmo trutta* L.). Journal of Fish Biology 4: 441–451.

Winfield, I. J. & N. C. Durie, 2004. Fish introductions and their management in the English Lake District. Fisheries Management and Ecology 11: 1–7.

Winfield, I. J., J. M. Fletcher & P. R. Cubby, 1998. The impact on the whitefish (*Coregonus lavaretus* (L.)) of reservoir operations at Haweswater, U.K. Archiv für Hydrobiologie, Official Journal of the International Association for Theoretical and Applied Limnology. Special Issues: Ergebnisse der Limnologie 50: 185–195.

Winfield, I. J., J. M. Fletcher & A. A. Lyle, 2008a. Assessment of the vendace refuge population of Loch Skeen. Final Report. Centre for Ecology & Hydrology Report to Scottish Natural Heritage. Scottish Natural Heritage, Edinburgh.

Winfield, I. J., J. M. Fletcher & J. B. James, 2008b. SCM of fish in standing waters 2007/2008 (Phase I). Final Report. Centre for Ecology & Hydrology Report to Scottish Natural Heritage. Scottish Natural Heritage, Edinburgh.

Winfield, I. J., J. M. Fletcher & J. B. James, 2008c. The Arctic charr (*Salvelinus alpinus*) populations of Windermere, U. K.: population trends associated with eutrophication, climate change and increased abundance of roach (*Rutilus rutilus*). Environmental Biology of Fishes 83: 25–35.

Winfield, I. J., J. M. Fletcher, J. B. James & C. W. Bean, 2009a. Assessment of fish populations in still waters using hydroacoustics and survey gill netting: experiences with Arctic charr (*Salvelinus alpinus*) in the UK. Fisheries Research 96: 30–38.

Winfield, I. J., J. M. Fletcher & J. B. James, 2009b. SCM of fish in standing waters 2007/2008 (Phase II). Final Report. Report to Scottish Natural Heritage. LA/C03380/6.

Winfield, I. J., J. M. Fletcher, J. B. James, C. W Bean & C. Duigan, in press. Setting reference values for lake fish abundance: observations from unexploited or lightly exploited vendace (*Coregonus albula*), whitefish (*C. lavaretus*) and Arctic charr (*Salvelinus alpinus*) populations in the U.K. In Cowx, I.G. (ed), Assessing the Ecological Status of Rivers, Lakes and Transitional Waters. Fishing News Books, Blackwell Scientific Publications, Oxford.

Wright, G. A., 2003. Impact of cormorants on the Loch Leven trout fishery and the effectiveness of shooting as mitigation. In Cowx, I. G. (ed.), Interactions between fish and birds: implications for management. Fishing News Books. Blackwell Scientific Publications, Oxford: 288–297.

Hydrobiologia (2012) 681:85–104
DOI 10.1007/s10750-011-0927-6

LOCH LEVEN RESEARCH

# Long-term variations in waterfowl populations in Loch Leven: identifying discontinuities between local and national trends

**D. Carss · B. M. Spears · L. Quinn · R. Cooper**

Published online: 4 November 2011


**Abstract** Loch Leven has been designated as a UK Ramsar Site (1976), a Site of Special Scientific Interest (1985) and a Special Protection Area (2000) due to its importance as a site for overwintering waterfowl. However, no comprehensive assessment of trends in waterfowl at the local versus national scale has been conducted at the site. Coherence between trends in 5-year mean species abundance for Loch Leven and Underhill Indexing Method values for Scotland (or GB in the case of geese) were assessed using principal components analysis for ten study species between 1968 and 2006. Five species showed trends at Loch Leven that were coherent with those at the Scottish scale (Eurasian Teal, Mute Swan, Great Cormorant, Pink-footed and Greylag geese). These species may not respond positively to local scale management. However, the other five species (Mallard, Coot, Great Crested Grebe, Tufted Duck and Pochard) showed distinct differences between local and national trends. A study of the feeding ecology for these species, the re-establishment of ringing effort and an assessment of waterfowl–wetland relationships are recommended to determine how changes in local food resources and habitat quality interact with macro-scale population dynamics to influence local and regional patterns of abundance.

**Keywords** Waterfowl · Loch Leven · Trend comparison · Local scale · National scale

Guest editors: L. May & B. M. Spears / Loch Leven: 40 years of scientific research

D. Carss · B. M. Spears (✉) · L. Quinn
Centre for Ecology & Hydrology, Penicuik, Midlothian EH26 OQB, Scotland, UK
e-mail: spear@ceh.ac.uk

R. Cooper
Scottish Natural Heritage, Battleby, Redgorton, Perth PH1 3EW, Scotland, UK

> "Lough Leven is four mile square and sixteen miles about. So far the land is good, but here and there many high rocks and hills: in this lough is fish every day gotten for store, none in Britain like, and consider the bigness of it as also for fowl. There is a river they call the Leven running out of it eight miles to the sea, and in it is salmons… there be great store of all kinds of wildfowl, of wild geese there being continually seen 3,000 or 4,000, and swans many."
>
> Sir Chistopher Lowther, 1629.

## Introduction

The sensitivity of particular waterfowl species to environmental change at various spatial scales may be driven by species-specific requirements (for examples, see Table 2). At national levels, migratory waterfowl may be influenced by the length and severity of winter weather, whilst those species that are reliant upon underwater macrophytes as a food source may decline during periods of intense eutrophication when macrophyte cover is reduced regionally (Boere et al., 2006). Similarly, those species that require stable shoreline habitat for breeding may decline locally as a result of wind-induced wave disturbance to riparian habitats (Rehfisch et al., 2004 [cit. Boere et al., 2006]).

Significant downward trends in waterfowl abundance of key species have been observed across a range of important stop-over and overwintering sites across Europe (e.g. the Netherlands—Lake Ijmeer, Lake Merkermeer, Lake Ijsselmeer; Russia/Estonia—Lake Pihkva (Van Eerden et al., 2007). The general response of lake managers to these downward trends has been to increase habitat complexity and quality in an attempt to increase ecosystem resilience to environmental pressures. This approach may only be useful for those species that are declining as a result of local level environmental pressures (e.g. smelt stock depletion in Lake Ijsselmeer). It is essential that trends and drivers be monitored and investigated if the conditions of the EU Birds Directive (79/409/EEC), Habitat Directive (92/43/EEC) and Water Framework Directive (2000/60/EC) are to be met in Europe's large shallow lakes.

Loch Leven has undergone significant environmental change in recent years, including eutrophication (Carvalho et al., 2011; May et al., 2011) and subsequent management (Bailey-Watts & Kirika, 1999; May & Spears, 2011), and climate change (Ferguson et al., 2008; Spears & Jones, 2010). Such changes are expected to alter waterfowl numbers at the species-specific level (Finlayson et al., 2006). However, it has been suggested that eutrophication effects, and those of associated management, should only act at the local scale, whilst climate change effects could be expected to act at local, national and even international scales (Boere et al., 2006). Although the international importance of Loch Leven as a site for waterfowl is well acknowledged, no comprehensive assessment of the effects of environmental change on the important waterfowl species has been conducted to date. Such an assessment is of particular importance given the conservation obligations associated with European legislation (e.g. Water Framework Directive and Habitats Directive).

A common approach in the assessment of the effects of environmental change on waterfowl species is to (1) identify trends in particular species at local and national scales, (2) identify discontinuities in local versus national trends and (3) where discontinuities are observed, identify the drivers of change at the site using available water quality and climate change data (Favero & Becker, 2006). Here, we use long-term bird monitoring data from Loch Leven to attempt steps (1) and (2), above, for some of Loch Leven's waterfowl species. We also explore the published literature to create hypotheses whereby step (3) could be undertaken in further studies.

## Methods

### Study site

Loch Leven (Fig. 1) is a shallow (mean depth = 3.9 m) lowland loch in east-central Scotland, UK. It has a perimeter of about 23 km and large expanses of shallow open water habitat. The comparatively large surface area, together with suitable islands for nesting, extensive shallow areas, rich adjacent farmland and relatively little disturbance, make Loch Leven a nationally and internationally important site for waterfowl (Allison & Newton, 1974). These authors examined the status of waterfowl at Loch Leven from 1966 to 1972, whilst more recent counts (until 1982) have been reported by Owen et al. (1986).

The significance of the loch as a site for overwintering waterfowl has been recognised through its designation as one of the original 13 Ramsar Sites in the United Kingdom in 1976, as a Site of Special Scientific Interest (SSSI) in 1985, and as a Special Protection Area (SPA) in 2000 (Scottish Natural Heritage, 2007). Some common qualifying features of these designations include (a) the support of over 20,000 wintering waterfowl, (b) internationally important numbers of pink-footed goose, Shoveler *Anas clypeata* and (c) nationally important wintering populations of several other wildfowl species (Great Cormorant *Phalacrocorax carbo*, Gadwall *A. strepera*, Teal *A. crecca*, Pochard *Aythya ferina*, Tufted Duck, Goldeneye *Bucephala clangula* and Greylag

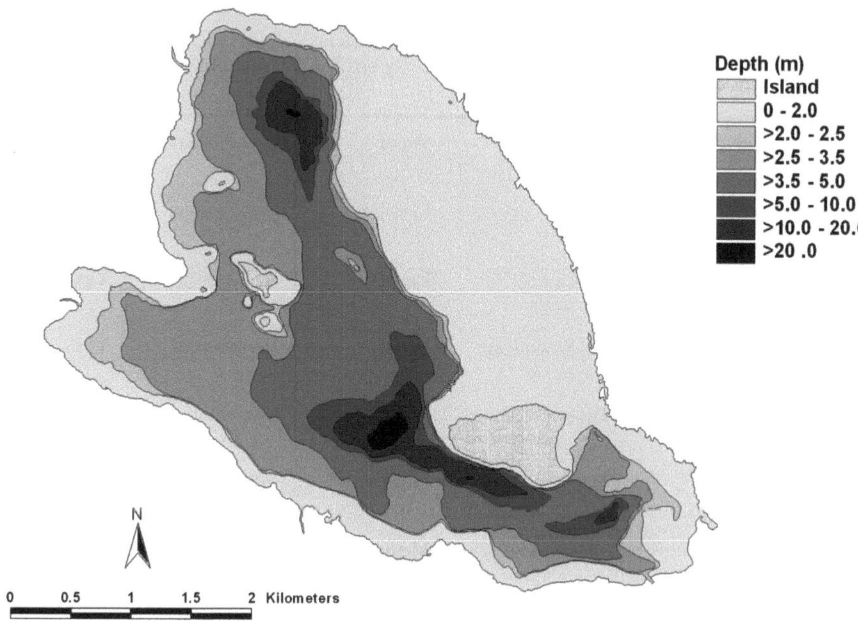


**Fig. 1** Bathymetric map of Loch Leven

goose (Scottish Natural Heritage, 2007)). Loch Leven was designated a National Nature Reserve (NNR) in 1964, with over 1,000 pairs of breeding duck (as such, it is one of the top three inland waters in Europe) and up to 35,000 birds recorded in autumn and winter. As such, it is the most important loch in Scotland for wintering/migratory wildfowl.

Loch Leven has a long and well-documented history of eutrophication (Carvalho et al., 2011; May et al., 2011) and recent investigations have highlighted its sensitivity to a range of climate change drivers, e.g. temperature (Ferguson et al., 2008; Carvalho et al., 2011) and wind (Spears & Jones, 2010; Spears et al., 2011). In relation to possible effects on bird habitats and food resources, long-term and seasonally distinct variations have been observed in the following variables: wind and wave intensity (Spears & Jones, 2010); macrophyte colonisation depth (May & Carvalho, 2010); macroinvertebrate community composition and biomass (Gunn et al., 2011), fish community composition and biomass (Winfield et al., 2011) and the occurrence of potentially harmful algal blooms (Carvalho et al., 2011). Whilst it is not yet possible to make direct links between long-term trends in waterfowl abundance and changes in any of these other parameters, possible links will be discussed, below, where appropriate.

## Bird counts

### *Choice of species*

Given the variety of waterfowl at Loch Leven, their different migration strategies and their various spatial and temporal uses of the loch and its surrounding land for breeding, feeding and loafing, ten species were selected for inclusion in this analysis (Table 1). These were categorised according to (a) their predominant feeding habit (i.e. surface feeding, diving to a variety of depths, or land-based), (b) their relative body size (i.e. small, medium or large) and (c) their predominant diet (i.e. omnivorous, herbivorous, fish-eater or invertebrate-eater). Furthermore, some individuals of some species are resident at Loch Leven throughout the year whilst others are predominantly passage or wintering birds there. For five of these species, Loch Leven is considered to be one of the most important Scottish sites in relation to breeding, moulting, or overwintering (Table 2).

### *Loch Leven data*

Autumn and winter (generally, September–March) waterfowl counts were compiled by Scottish Natural Heritage (SNH) based on surveys carried out from

**Table 1** Details of the ten waterfowl species included in this study

| Species | Predominant feeding habit | Body size[a] | Predoninant diet | Present all year? | Mainly winter visitor? | L. Leven a main site? |
|---|---|---|---|---|---|---|
| Eurasian Teal *Anas crecca* | Surface | Small | Omnivorous | | Yes | |
| Mallard *Anas platyrhynchos* | Surface | Medium | Onmivorous | Yes | | |
| Mute Swan *Cygnus olor* | Surface | Large | Herbivorous | Yes | | |
| Coot *Fulica atra* | Shallow diver | Small | Herbivorous | Yes | | Yes |
| Great Crested Grebe *Podiceps cristatus* | Diver | Medium | Fish-eater | Yes | | Yes |
| Great Cormorant *Phalacrocorax carbo* | Diver | Large | Fish-eater | | Yes | |
| Tufted Duck *Aythya fuligula* | Benthic diver | Medium | Invertebrates | Yes | | Yes |
| Common Pochard *Aythya ferina* | Benthic diver | Medium | Omnivorous | Yes | | |
| Pink-footed goose *Anser brachyrhynchus* | On land | Large | Herbivorous | | Yes | Yes |
| Greylag goose *Anser anser* | On land | Large | Herbivorous | | Yes | Yes |

Information from Forrester & Andrews (2007) unless otherwise stated ([a] Bruun & Singer, 1978)

land-based vantage points between 1968 and 2006. Counts were conducted by SNH at monthly intervals between 1967 and 1994, and at monthly intervals using the Wetland Bird Survey methods (Austin et al., 2008) between 1995 and 2003. Since then, fortnightly surveys have been carried out by SNH in collaboration with the Royal Society for the Protection of Birds (RSPB, Vane Farm, Kinross, Scotland KY13 9LX, UK). Although the peak months for each bird species may vary between September and March, we have chosen to use mean values across this period as a comparable index of change in mean winter abundances. Available data were processed to produce 5-year running mean winter average abundances for the ten species under investigation.

*National trend data*

National data were obtained from the British Trust for Ornithology (BTO). Scottish trend data were available for all of the study species with the exception of the two goose species. GB trend data were provided for these. Trends were calculated by the BTO using the Underhill indexing method (UIM) and are discussed in detail by Underhill (1989) and Underhill & Prŷs-Jones (1994). The UIM uses a multiplicative bootstrap modelling approach to fill data gaps and create a continuous series of bird count estimates with which an index number can be produced. The index number expresses the change in population size in a given year in relation to a base year as a ratio. The main assumptions used in the UIM are that: (1) variations in site, year and month are independent of each other, (2) site factors remain constant with time and (3) the effects of month and year are equal for all sites. A critical comparison of the advantages and limitations of the UIM and other indices is provided by Kirby et al. (1995) and Underhill & Prŷs-Jones (1994). UIM values were processed to produce 5-year mean values corresponding to the species abundance numbers that were produced separately from the Loch Leven data.

**Table 2** Summary of key information for the ten waterfowl species included in this study

| Species | Feeding behaviour[a] (unless otherwise stated) | Diet | Habitat requirements | Population | Presence at Loch Leven | Range of Loch Leven data |
|---|---|---|---|---|---|---|
| Eurasian Teal *Anas crecca* | Surface feeding "dabbler", also filters mud whilst walking in shallow waters, up-ends and picks items off surface of water. Mostly nocturnal | Small invertebrates sieved from water's surface, winter diet mainly seeds of grasses/sedges and aquatic plants. Also forages on land, e.g. autumn stubble fields | Wide variety of waterbodies used but oligotrophic pools favoured. Thus, upland peatlands more commonly used than lowland eutrophic waters | Breeding pairs in Scotland: 1,950–3,400<br>Winter population in Scotland: c. 37,500 (22,500–125,000)<br>Loch Leven: 10–22 pairs but in decline since 1995 | Mainly a winter visitor, although up to 500 birds recorded moulting by mid-July | 1968 to present |
| Mallard *Anas platyrhynchos* | Surface feeder—dabbling and up-ending, but also diving. Also takes food on land, plucking leaves and shoots and grazing like geese | Omnivorous: insects, molluscs, tadpoles, crustaceans plus the seeds, leaves and stems of aquatic plants | Fresh, brackish and salt water, from temporary water bodies to large lochs, still or flowing waters | Breeding pairs in Scotland: 17,000–43,000<br>Winter population in Scotland: 65,000–90,000<br>Loch Leven: average breeding pairs over 400 | Resident; may be augmented by influxes from elsewhere in Scotland and by small numbers from abroad during cold spells | 1968 to present |
| Mute Swan *Cygnus olor* | Surface feeder, up-ending and dipping | Herbivorous: submerged (up to 1 m) and emergent vegetation, also pasture and sometimes cereals sown close to water edge[a] | Still waters, canals and the lower reaches of rivers. Estuaries, sheltered marine shores | Territorial pairs in Scotland: 1,375 (1,012 of which breed)<br>Winter population in Scotland: 5,000–8,000<br>Loch Leven: 10+ breeding pairs | Resident: may relinquish territories in freezing conditions<br>Larger numbers may arrive overwinter | 1968 to present |
| Common Coot *Fulica atra* | Diver: 1–2 m depth, up-ends in shallower water, and feeds from surface[b] | Omnivorous: mainly vegetative parts and seeds of aquatic plants and grass, invertebrates and small fish taken where available[b] | Nutrient-rich lochs, ponds, canals and rivers with good growth of emergent and submerged vegetation. In winter, flocks congregate on large lowland waters | Breeding pairs in Scotland: 1,500–2,500<br>Winter population in Scotland: 6,000–11,000 birds<br>Loch Leven: 100 breeding pairs in late 20th Century, reduced to 25 pairs by 2004 | Important breeding site in Scotland and primary moulting site in Scotland with 1,000+ birds recorded by midsummer. Holds unknown numbers of passage migrants and overwintering birds from Northern Europe | 1968 to present |

**Table 2** continued

| Species | Feeding behaviour[a] (unless otherwise stated) | Diet | Habitat requirements | Population | Presence at Loch Leven | Range of Loch Leven data |
|---|---|---|---|---|---|---|
| Great Cormorant *Phalacrocorax carbo* | Pursuit-diver; can dive up to 30 m but 3–9 m most common. Catches fish from the water column and after from the bottom | Piscivorous: almost exclusively fish, mostly individuals less than 25 cm. A "generalist" and opportunist predator taking highly diverse range of prey | Forages in all but the smallest water bodies, including rivers, canals, lakes, ponds, estuaries and shallow coastal waters | Breeding pairs in Scotland: c. 3,600<br>Winter population in Scotland: 9,000–15,000<br>Loch Leven: mean of 379 birds wintering on Loch Leven 1999–2004 | Some presence year round. Considerable numbers overwinter at the loch. Turnover is high as birds are mobile, regularly moving to neighbouring foraging sites and also farther afield | 1981 to present |
| Tufted Duck *Aythya fuligula* | Benthic diver, sustained dives to 3–14 m, usually 3–5 m in breeding season. Feeds mainly on the bottom Also up-ends and forages by wading in the shallows | Wide range of invertebrates, usually molluscs but varies with locality and period. More than 80% of diet at Loch Leven is Chironomids[c] | Shallow, open water bodies with invertebrate-rich benthic sediments and submerged vegetation. Nests in tall, rank herbaceous cover usually within 100 m or shore and often on islands | Breeding Pairs in Scotland: 2,250–2,700<br>Winter Population in Scotland: 11,000<br>Loch Leven: 200–550 breeding pairs (2006) | Resident (Loch Leven one of the most important breeding sites in Britain), usually only low numbers in winter | 1968 to present |
| Common Pochard *Aythya ferina* | Benthic diving duck: up to 3 m in open water uncluttered by floating vegetation | Omnivorous, aquatic plant material includes seeds, rhizomes, buds, shoots, leaves, tubers. Sedges and grasses also eaten. Crustaceans, molluks annelids, insects and larvae, especially Chironomids[a] | Lowland, nutrient-rich, eutrophic open waters, with dense submerged plant growth & abundant invertebrate prey | Breeding pairs in Scotland: 25–50<br>Winter population in Scotland: 4,000–6,000 birds<br>Loch Leven: 5–14 pairs in 2004 | June–August to moult, others winter on the loch from September–February. Most birds return to the continent to breed | 1968 to present |

**Table 2** continued

| Species | Feeding behaviour[a] (unless otherwise stated) | Diet | Habitat requirements | Population | Presence at Loch Leven | Range of Loch Leven data |
|---|---|---|---|---|---|---|
| Pink-footed goose *Anser* | *brachyrhynchus* | Herbivorous forager taking accessible vegetation from ground. Most feeding takes place on farmland, mainly by day | In Scotland, feeding almost exclusively on farmland: on grass, arable and root crops, barley stubble and winter cereals[a] | In Scotland in winter, almost exclusively found foraging on farmland and generally roosting on water | Winter population in Scotland: 200,000 (October), 100,000–150,000 (winter/spring). Representing roughly half of the world's population | |
| Loch Leven: Regular midwinter numbers of around 6,000 | Overwintering immigrants from Iceland | | | | 1968 to present | |
| Greylag goose *Anser anser* | Herbivorous forager taking accessible vegetation from ground or water surface | Wide variety of aquatic plants, agricultural grasses and sometimes root crops (summer), Grass, cereal grains, root crops and growing cereals (winter)[a] | Breeds amongst tall vegetation, around shores of lochs, amongst peat bogs, in deep heather on moors, along riversides alongside areas of permanent pasture. Forages in marshes, lake maring and low-lying farmland, including pasture | Breeding population in Scotland: 20,000 'native' birds, 5,000 'naturalised' birds (at least 700 pairs)<br>Winter (Scotland): Excess of 85,000 wintering birds plus resident breeding populations<br>Loch Leven: 150+ pairs | Resident plus large numbers of overwintering immigrants from Iceland | 1968 to present |

Information from Forrester & Andrews (2007) unless otherwise stated ([a] Cramp & Simmons, 1977; [b] Cramp & Simmons, 1979; [c] Stewart & Lauder, 1997)

## Statistical analysis

Trends in 5-year mean species abundance for Loch Leven and UIM values for Scotland (or GB in the case of geese) were assessed visually for all ten study species. Principal Components Analysis (PCA) using correlation was used to produce the two synthetic axes that best captured the variation in (i) the 5-year means for each of the species and (ii) trends across Loch Leven and Scotland. The exceptions to this were greylag geese and pink-footed geese, where trends across Loch Leven and GB were included in the analysis. The distributions of each variable across the first two principal components were used to highlight groups of bird species that varied similarly across each spatial scale (i.e. Loch Leven vs. Scotland/GB). All statistical analyses were carried out using Minitab statistical software, version 14 (Minitab Ltd., Coventry, UK).

## Limitations of data analysis

Whilst several of the waterfowl species considered here are present at Loch Leven throughout the year and breed there (often in large numbers), the present analysis was restricted to winter counts only. Birds were counted on the loch in each of the 7 months between September and March each year and the final figure for each winter was taken as the largest of these monthly counts. Such winter counts were used primarily to standardise, as much as possible, the methods used to count birds and thus allow direct comparisons between species. However, it also meant that we had access to runs of data for the Loch's waterfowl spanning four decades. Whilst there were many advantages in focusing on winter counts, their use raised a major issue throughout this article in relation to the numbers of birds counted in the winter: the relationship between local phenomena (e.g. the influence of local breeding productivity) and wider—national or international—ones (e.g. the influence of continental immigration).

In order to visualise patterns in the data more clearly, 5-year rolling means were calculated for each species, each annual figure being the mean value for a 5-year period covering the year in question and the preceding four winters. Whilst this method allowed us to draw out any clear temporal trends in the data, it did mean that the precise timing of some events (e.g. the start/end of a decline/increase/stable period) would not be clear in the graphs produced even if they were apparent in the annual raw counts data. Thus, rather than discussing events on an annual basis, we have described them on a five-year or even decadal level.

# Results

## Comparing local and national long-term trends

In each of the sub-figures presented here, five-year rolling means are shown graphically. Each point on the graph represents the average of the five preceding winter counts. Thus, a point shown as '1995' on the $x$-axis is the average of winter bird counts for 1990/1, 1991/2, 1992/3, 1993/4 and 1994/5.

### Surface feeders

#### *Teal*

Average 5-year mean winter Teal numbers at Loch Leven (Fig. 2a) were relatively constant at around 600 birds between the early-1970s and the early-1980s. After this, there was a general and steady increase in Teal at the site through to the mid-2000s when numbers reached around 5,000. Average counts in more recent years indicate a decline in numbers of, perhaps, around 500–600 birds. The trend observed in Loch Leven was very similar to that across Scotland, with many peaks and troughs synchronised, and the post-2005 decline being clear at both spatial scales.

#### *Mallard*

Average 5-year mean winter Mallard numbers at Loch Leven (Fig. 2b) declined from over 2,500 birds to 1,600 between the early- and the late-1970s and then increased to around 2,500 again by the mid-1980s. Thereafter, numbers declined again to around 1,400 birds by the mid-1990s and have since remained relatively stable (at around 1,500 birds). A similar general trend was observed at the Scottish scale, although trends have diverged markedly since the mid-1990s when Mallard numbers at Loch Leven fluctuated around a steady mean while the trend across Scotland continued a lengthy and steady decline.

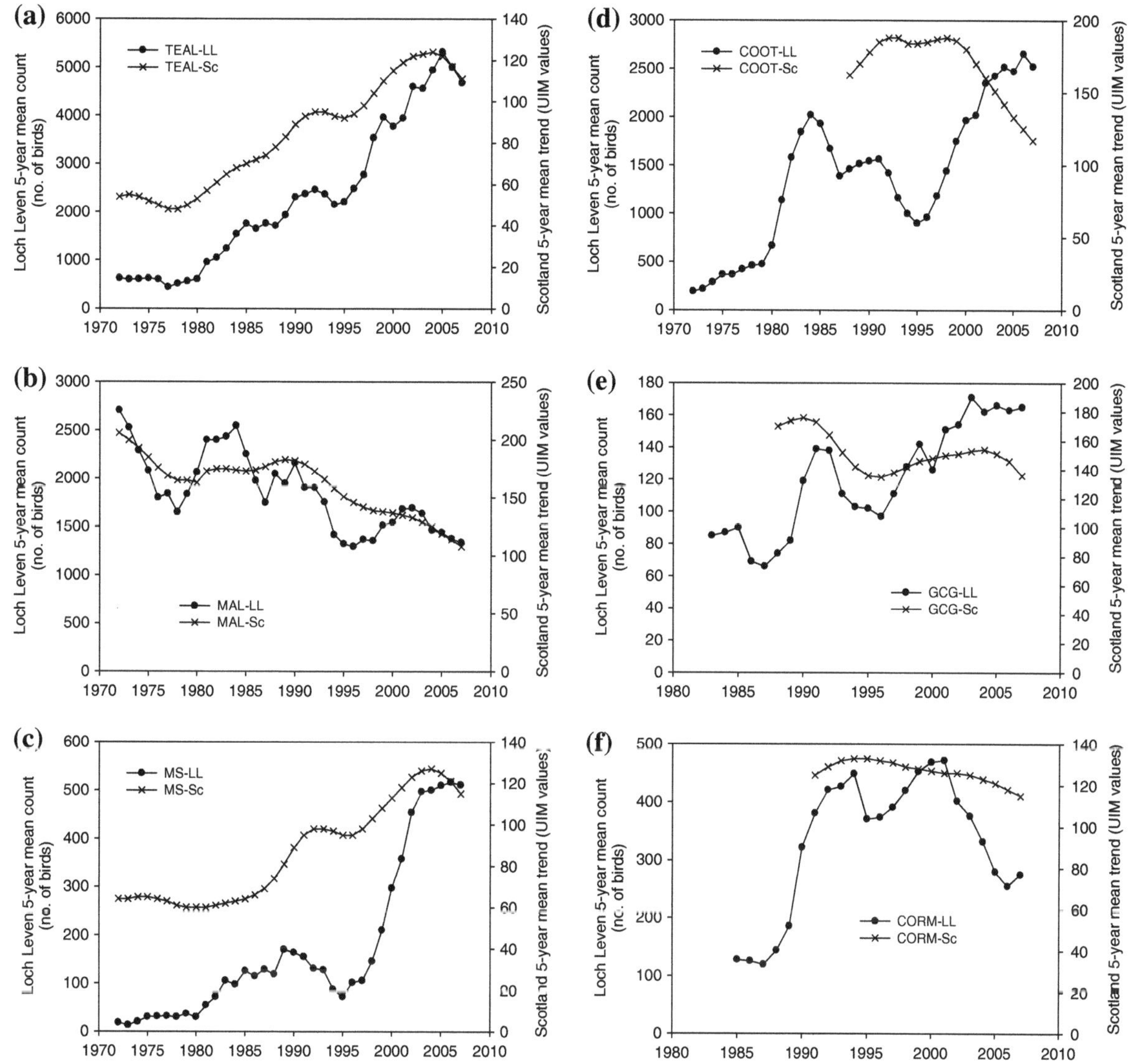


**Fig. 2** Comparison of 5-year mean winter trends for **a** Teal (*TEAL*), **b** Mallard (*MAL*), **c** Mute Swan (*MS*), **d** Coot (*COOT*), **e** Great Crested Grebe (*GCG*), **f** Cormorant (*CORM*), **g** Tufted Duck (*TUFT*), **h** Pochard (*POCH*), **i** Pink-footed goose (*PFG*) and **j** Greylag goose (*GREY*) at Loch Leven and across either Scotland (**a–h**) or Great Britain (**i, j**). *UIM* underhill indexing method

### *Mute Swan*

Average 5-year mean winter Mute Swan numbers at Loch Leven (Fig. 2c) were relatively constant at fewer than 50 birds between the early-1970s and the early-1980s. Thereafter, numbers increased to around 150 birds in the late-1980s before declining to about 100 birds around 1995. A general, and sharp, increase in numbers was recorded in the next decade to around 500 birds and, after the mid-2000s, Mute Swan numbers at Loch Leven increased comparatively slowly, perhaps indicating an onset of stabilisation. The trends observed in Loch Leven were broadly coherent with those across Scotland, except for the recent marked decline in the Scottish trend during which numbers at Loch Leven continued to increase steadily.

## Various divers

### *Coot*

Average 5-year mean winter Coot numbers at Loch Leven (Fig. 2d) increased steadily to around 500 birds between the early-1970s and the end of that decade,

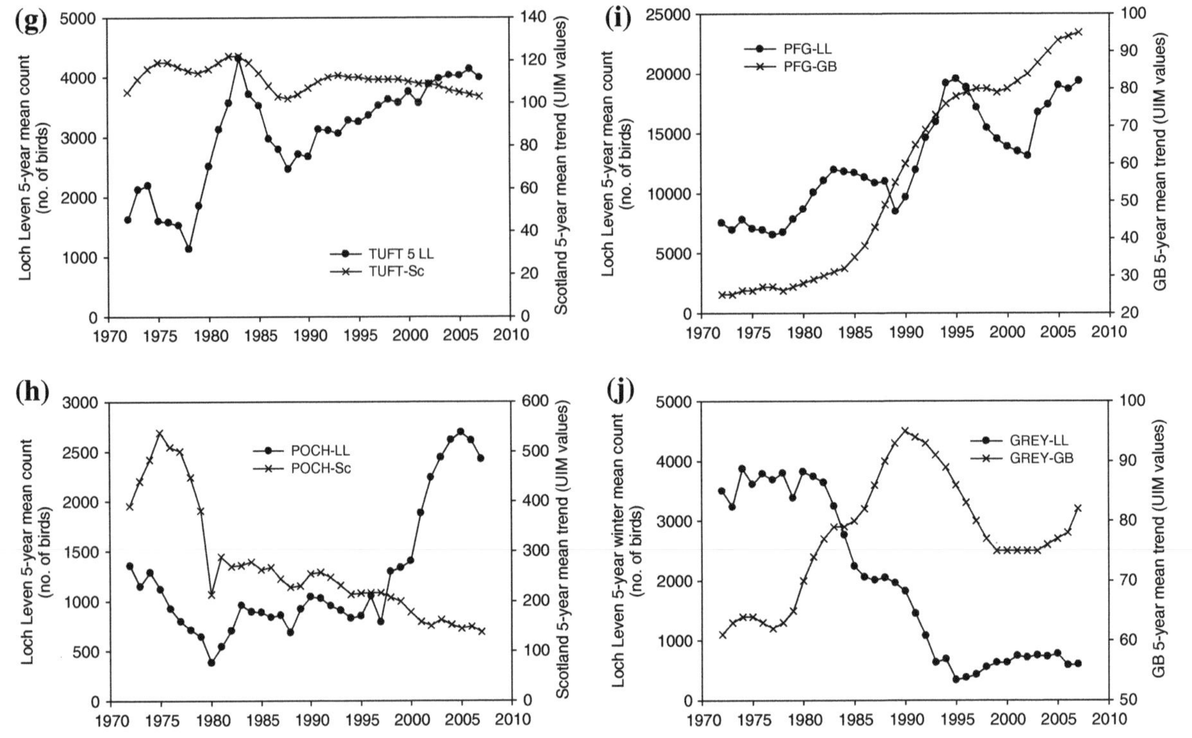


**Fig. 2** continued

with a sharp increase to around 2,000 birds in the mid-1980s. Numbers then generally declined to around 1,000 in the mid-1990s and, thereafter, numbers increased relatively sharply to around 2,500 birds in the mid-2000s. No coherence was observed between the Loch Leven and Scottish trends. Scottish trends showed an increase between the mid-1980s and mid-1990s, and a general stabilisation until the late-1990s. After that, Scottish numbers declined dramatically, unlike those at Loch Leven which have increased since then.

*Great Crested Grebe*

Average 5-year mean winter Great Crested Grebe numbers at Loch Leven (Fig. 2h) declined from around 90 to about 70 birds between the early-1970s and the late-1980s. This was followed by a sharp increase to 140 birds by the early-1990s, decrease until the mid-1990s, and then a steady increase to around 160 birds by the early-2000s. Since then, numbers appeared to stabilise. The Scottish trend was similar to that at Loch Leven in that the decline and subsequent increase after the mid-1990s was similar. However, whilst the subsequent trend at Loch Leven appeared to be stabilising, that at the Scottish level showed a dramatic decline after the early-2000s.

*Great Cormorant*

Average 5-year mean winter cormorant numbers at Loch Leven (Fig. 2f) increased from around 120–450 birds between the mid-1980s and the early-1990s and, following a slight decrease, numbers continued to rise to around 480 birds in the early-2000s. After this, there was a sharp decline to around 250 birds in the mid-2000s. With the exception of the Loch Leven peak in the early-2000s, the local trend showed a lagged coherence with that for Scotland as a whole.

*Tufted Duck*

Average 5-year mean winter Tufted Duck numbers at Loch Leven (Fig. 2g) showed a general decline from around 2,000 to around 1,100 birds between the early-1970s and the late-1970s. Numbers then increased sharply to around 4,500 birds towards the early-1980s before declining just as sharply to around 2,500 birds in the late-1980s. Thereafter, there was a steady, but less dramatic, increase to around 4,000 birds in recent years. The Loch Leven trend is broadly similar to that at the Scottish scale, especially in relation to the early-1980s peak and late-1980s trough. However, there has been a marked divergence since the early-1990s, with the Scottish trend showing a steadily decline whilst the Loch Leven trend has continued to increase.

*Pochard*

Average 5-year mean winter Pochard numbers at Loch Leven (Fig. 2h) declined from around 1,400 birds in the early-1970s to around 500 birds by about 1980. Numbers then increased to around 1,000 birds in the early-1980s and remained relatively stable until the late-1990s. A sharp increase followed to around 2,700 birds by the mid-2000s and, since then, numbers have declined again to around 2,400 birds. Coherence between the Loch Leven and Scottish-scale trends was slight and seemed to occur only between the late-1970s and the late-1990s. After this, the Scottish trend continued to decline whilst that at Loch Leven increased sharply.

Land-based foragers

*Pink-footed goose*

Average 5-year mean winter pink-footed goose numbers at Loch Leven (Fig. 2i) increased from around 7,500 to 12,000 birds between the early-1970s and early-1980s. After this, they declined to around 9,000 birds in the late-1980s. There was a sharp increase, thereafter, to 20,000 birds by the mid-1990s, followed by another decline, to around 14,000 birds, in the early-2000s. Numbers then increased again to about 20,000 birds between the early-2000s and 2007. The GB trend increased steadily throughout the monitoring period and appeared to be broadly coherent with the Loch Leven trend, albeit without the temporary periods of decline.

*Greylag goose*

Average 5-year mean winter greylag goose numbers at Loch Leven (Fig. 2j) remained steady at 3,000–4,000 birds during the 1970s before declining to less than 500 birds by the mid-1990s. From then onwards, numbers increased very gradually towards 1,000 birds by the mid-2000s with a slight decrease evident towards 2007. The GB trend appeared to be broadly coherent with that for Loch Leven, at least between 1990 and 2005, although in recent years an increasing trend has been observed. Prior to 1990, when Loch Leven numbers were either stable or declining, there was a reasonably steady increase at the GB-scale.

**Table 3** The eigenvalues (*EV*), proportion (*P*) and cumulative proportions (*CP*) of the variance explained by principal components 1 (PC1) and 2 (PC2)

| | *EV* | *P* | *CP* |
|---|---|---|---|
| PC1 | 11.81 | 0.738 | – |
| PC2 | 0.894 | 0.156 | 0.894 |

Assessment of covariation in species trends across local and national scales

The PCA analysis (Table 3) identified three distinct groups of covariation (Fig. 3). The majority of trends in bird species at Loch Leven were explained either through positive or negative association with PC1, with the exception of Great Crested Grebe (Scottish trend) and Mallard (Loch Leven trend), which were strongly negatively associated with PC2 (Table 3). The analysis highlights a number of discrepancies in relation to covariation between species and across trend scales. Strong covariation was indicated between Loch Leven and Scottish trends for Teal and Mute Swan only. For cormorants, the Loch Leven and Scottish trends were positively related to PC1 and appeared to weakly covary with each other. Loch Leven and Scottish trends of Tufted Duck, Pochard, and Coot all showed contrasting variation along PC1. The results of the PCA are in good agreement with the visual analysis.

## Discussion

Comparison of Loch Leven versus national and international trends

Of the ten species examined in this paper, five showed trends at Loch Leven that were coherent, or broadly so,

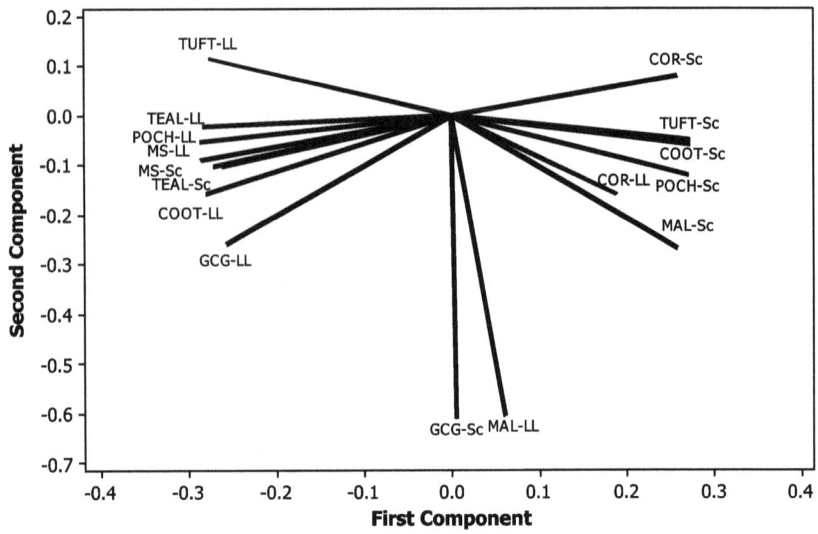


**Fig. 3** Results of principal components analysis showing the ordination of the national (i.e. Scotland—SC) and local (Loch Leven—LL) trends in the eight bird species along principal components 1 (first component) and 2 (second component). *TEAL* Teal, *MAL* Mallard, *MS* Mute Swan, *COOT* Coot, *GCG* Great Crested Grebe, *COR* Cormorant, *TUFT* Tufted Duck, *POCH* Pochard

with the trend at the Scottish scale. These were two surface feeders (Eurasian Teal and Mute Swan), one diver (the Great Cormorant) and both goose species (pink-footed and greylag). The other five species all showed distinct differences between the local Loch Leven trends and those at the Scottish scale. These comprised a surface feeder (Mallard) and four of the five diving species (Coot, Great Crested Grebe, Tufted Duck, Pochard). Whilst the primary focus of this study was to examine instances where local trends at Loch Leven differed substantially from those at the national (i.e. Scottish) level, it was also useful to briefly review those cases where local and national trends were similar. Furthermore, it was of interest to put the local and national pictures into a wider European context. Whilst it was not possible to get directly comparable data (i.e. standardised monthly winter counts) at this international scale, comparisons were drawn between the Loch Leven and national (e.g. Scottish or GB trends) situations and recent data on European breeding population trends (1970–2000) reviewed by Burfield & van Bommel (2004). This latter assessment forming the basis for identifying species of conservation concern in relation to both EU Birds Directive (79/409/EEC) and Habitat Directive (92/43/EEC).

## Similar local versus national trends

### *Eurasian Teal*

Whilst breeding in Scotland, the Eurasian Teal is more commonly a winter visitor and passage migrant from northern Europe and Fennoscandia (Forrester & Andrews, 2007, p. 195). Both Scottish and Loch Leven trends in the wintering population of Eurasian Teal have increased overall since the 1960s, in line with trends for the UK as a whole. Whilst the reasons for this increase are unclear, they may relate to increased annual survival rates, migratory short-stopping or responses to such factors as milder winters and/or reduced shooting pressure (Forrester & Andrews, 2007, p. 199). These increases are not necessarily reflected at the European scale (Burfield & van Bommel, 2004) where the large breeding population of Teal (over 920,000 pairs) was stable between 1970 and 1990 and, whilst declining in some countries during 1990–2000, probably declined slightly overall.

### *Mute Swan*

The Mute Swan is a widespread resident breeding bird in Scotland. Movements within Scotland appear to be fairly localised and there have been no records of Scottish-ringed birds being recovered abroad and only two birds from abroad have ever been recorded in Scotland (Forrester & Andrews, 2007, pp. 128–129). Thus, the increased overall Scottish and Loch Leven trends in the wintering population of Mute Swan since the 1960s are thought to be the result of a long series of mild winters and associated lower mortality rates and increases in juvenile survival (Forrester & Andrews, 2007, p. 130). These increases are similarly reflected at the European scale (Burfield & van Bommel, 2004) where the relatively small breeding population of

Mute Swan (less than 120,000 pairs) increased moderately between 1970 and 1990 and, whilst declining in some countries during 1990–2000, most European populations (including key ones in the UK) increased or were stable. As such, the European population has undergone a large increase overall.

*Great Cormorant*

The Great Cormorant is a relatively scarce coastal breeder in Scotland. It is distributed much more widely outside the breeding season, both around the coasts and in many major river catchments and larger lochs. Movements at this time of year can be extensive with birds moving both within the UK and between the UK and the continent (Forrester & Andrews, 2007, p. 400). Whilst the Scottish winter population of Great Cormorant has remained relatively stable in recent years (Scottish Natural Heritage, 2009), that at Loch Leven has fluctuated more dramatically. At the European scale (Burfield & van Bommel, 2004), the large breeding population of Great Cormorant (over 310,000 pairs) increased greatly between 1970 and 1990 and continued to do so during 1990–2000 with almost all national trends being either stable or increasing.

Clearly, site-specific issues, particularly food availability, affect the local abundance of Great Cormorants and as these birds are considered by many to have negative effects on recreational fisheries (Carss et al., 1997; Carss & Marzano, 2005; Winfield et al., 2011) their presence at Loch Leven has generated much interest and a few salient points are discussed here.

At Loch Leven, average cormorant numbers were relatively low until the winter of 1989/90 then almost trebled by the winter of 1994/95 (Scottish Natural Heritage, 2009), a level of increase that was common to many inland locations across the UK at that time (Carss et al., 1997). Thereafter, numbers peaked in the winter of 2000/01 before beginning a steady decline (Scottish Natural Heritage, 2009). Whilst it is not possible to quantify the impact of Great Cormorants on the Loch Leven trout fishery (see Carss et al., 1997; Winfield et al., 2011 and references therein), it is likely that the presence of cormorant, there, has been influenced by both 'natural' changes in the fish community and by fisheries management strategies.

Historical information on Great Cormorant diet at Loch Leven was reviewed by Carss et al. (1997). There was a marked shift from a perch-dominated diet (by numbers and mass) during 1972–1977, through a period when the proportions of perch in the diet were declining and those of trout were increasing (1981–1991), to a trout-dominated diet (again, by numbers and mass) by 1992–1994. Great Cormorants are generalist predators, usually consuming the most abundant fishes or those that are most easy to catch at a particular foraging location (Carss & Marquiss, 1997). In general, the composition of the prey taken by the birds, and that present in the foraging location, are fairly similar and so the food composition of the diet can give a reasonable picture of the fish stock and its composition in the foraging waters (Hald-Mortensen, 1995; Carss & Marquiss, 1997). Thus, the recorded decline of perch in the diet of Great Cormorants at Loch Leven most probably reflects a reduction in their abundance in the loch, itself.

The perch at Loch Leven have been subject to an epizootic infection, the so-called 'perch disease', known to dramatically increase mortality of all age groups (Bucke et al., 2006). Whilst precise dates for this outbreak are uncertain, it was certainly an issue at the loch in the late-1970s and early-1980s when many dead fish were observed (May, pers comm.). Mortality from this disease can be very high, for example over 98% of the adult perch in Windermere were estimated to have died from it during 1976 (Bucke et al., 2006). Whilst the effects of the disease on Loch Leven's perch population are unknown, it is likely that it was a major source of mortality there.

As described by Winfield et al. (2011), the perch population of Loch Leven is also likely to have declined markedly as a result of nutrient enrichment, which began in the early-1970s, and an associated reduction in macrophytes in the loch. By the mid-1990s, nutrient input to the loch had declined substantially (May et al., 2011) and with the resulting ecological recovery in water quality (Carvalho et al., 2011), macrophytes (May & Carvalho, 2010) and invertebrates (Gunn et al., 2011), the perch population may now also be expected to show some recovery (Winfield et al., 2011). If this is the case, perch may once again become an important component of the Great Cormorant's diet at the loch.

Indeed this apparent shift in trophic web structure to a dominance of submerged macrophytes may be similar to that reported in other shallow eutrophic lakes. For example, Hargeby et al. (1994) report switches between two stable states as a result of an interaction between abundant macrophytes and increased water transparency. In turn, the abundance of macrophytes has a favourable effect on the recruitment of perch in a stabilised feedback loop (Hargeby et al., 1994). Moreover, as such a trophic state has also been shown to benefit a range of waterfowl including herbivores, diving ducks and fish-eaters, it is perceivable that the decline and recovery of the macrophyte population (Dudley et al., 2011) may be behind at least some of the local changes in numbers of the other waterfowl species at Loch Leven.

As described elsewhere (Winfield et al., 2011) stocking of trout directly into the loch began in 1983 in response to a dramatic decline in recreational fishery performance. Most of these stocked trout were in the size range most commonly taken by cormorants (cf. Cramp & Simmons, 1977), although evidence suggested that Great Cormorants were often selecting larger fish (Carss et al., 1997). Subsequently, in 2004–2006, older and larger fish were stocked. These were of a size that could be taken by anglers, but were also within the most commonly recorded size range of trout found to be taken by Great Cormorants at the loch in earlier surveys, i.e. 1972–1977 and 1992–1994 (Carss et al., 1997). Between 1983 and 1994, over 1.3 million trout were released to the loch (Carss et al., 1997) and stocking of in excess of 100,000 fish per year continued until 2004. Then, stocking was drastically reduced before being discontinued in 2006 (Winfield et al., 2011). These authors suggested that fish stocking at Loch Leven may have attracted more Great Cormorants to the site. In addition, our data suggest that, with the cessation of such intensive fish stocking at Loch Leven, the number of overwintering Great Cormorants visiting the site may have started to decline. Whether or not the end point reaches levels last seen in the 1980s, or even earlier, remains to be seen.

*Pink-footed goose*

For the pink-footed goose populations of Iceland and Greenland, Scotland is a key wintering area and there are large feeding and roosting flocks in the eastern and central regions (Forrester & Andrews, 2007, p. 141). The massive increase in the wintering population of pink-footed geese at both Loch Leven and at the Scottish scale has been attributed to a combination of factors, including increased areas of winter cereals and permanent grass, a gradual decline in shooting pressure and protection of many of the major roosts. (Forrester & Andrews, 2007, p. 143). Loch Leven is, primarily, a stop-over site for pink-footed geese, the vast majority of which continue their migration southwards to Norfolk and Lancashire. The geese do not use the loch directly for food; instead they feed on the surrounding catchment, split grain/arable stubble and root crops, during autumn and then favour new growth arable and pasture as the season progresses. Thus, goose abundance in particular locations, and at particular times, is strongly affected by the availability of local food resources. Indeed, recent agricultural changes in Scotland (and elsewhere in the UK), involving very early autumn cultivation and an associated reduction in stubble feeding opportunities, have been reflected in distribution patterns of wintering pink-footed geese (Thom, 1986, p. 97), and hence local, site-specific numbers. The local and national increases shown in this study are similarly reflected at the European scale (Burfield & van Bommel, 2004) where the relatively small breeding population of pink-footed goose (less than 69,000 pairs) increased substantially between 1970 and 1990 and continued to do so during 1990–2000, the species undergoing a moderate increase overall.

*Greylag goose*

In Scotland, the greylag goose is a common resident with two populations; a 'native' one in the north and west and a 'naturalised' one in the south and east. Whilst Scottish birds are known to migrate within the UK, the biggest movements are outside the breeding season when birds are joined by a large number of Icelandic immigrants (thought to be about 95% of this breeding population), which winter in lowland areas of Scotland (Forrester & Andrews, 2007, p. 151). Both of the Scottish based populations, i.e. native and naturalised, have shown a rapid increase in numbers, with a combined total estimated to be around 25,000 birds after breeding. Conversely, the wintering Icelandic population that increased up to 1990 has since shown a decline, numbering some 83,500 birds in late-2004. Goose distribution (and hence local numbers) is associated with the availability of appropriate agricultural land, such as spillage grain on stubbles in autumn, root crops and grass later in winter, and turnips and young grass and corn in spring (Thom, 1986, p. 103). Importantly, there have been major distribution changes in Scotland since the 1960s—a general redistribution northwards from the Lothian and Borders area in the south east towards the north-east of Scotland during the 1980–1990s and then up to Orkney during the 1990–2000s (Forrester & Andrews, 2007, p. 154). This northerly redistribution is likely to have caused the dramatic decline in greylag goose numbers observed at Loch Leven. Here, the species is one of the factors that qualify the site as a SSSI and there has been a noticeable population decrease since the mid-1980s as the geese redistribute themselves further northward. The greylag goose is thus an interesting example of a management quandary—a species whose numbers are decreasing at a site for which they are internationally designated while changes in distribution, especially of the more northerly wintering of Icelandic birds, may, in reality, be due to changes that cannot be fully compensated for by management measures implemented at the local scale. Despite a northerly redistribution of birds formerly wintering at Loch Leven, the rapid national increase shown in the present study is similarly reflected at the European scale (Burfield & van Bommel, 2004) where the large breeding population of greylag goose (over 120,000 pairs) increased substantially between 1970 and 1990. Although key Icelandic and Russian populations declined during 1990–2000, many other populations in Europe (incluing the UK) increased and so the species increased overall.

Dissimilar local versus national trends

*Mallard*

Although the western European (including British) populations of Mallard are mostly sedentary (Forrester & Andrews, 2007, p. 202), birds from higher latitudes arrive in Scotland from mid-September onwards. Wintering flocks may also contain birds that are moving around Scotland, together with small numbers of continental birds during cold periods (Forrester & Andrews, 2007, p. 203). Since the mid-1990s, numbers have declined at the Scottish scale (Fig. 2b). Although the reasons for this winter decline in numbers are unclear, reduced immigration could be expected during milder winters in continental and northern Europe. In addition, estuaries and coastal waters are now less polluted and the availability of invertebrate prey may have reduced as a consequence (Forrester & Andrews, 2007, p. 205). During the same time period, and in sharp contrast to the picture across Scotland, Mallard numbers at Loch Leven have remained relatively stable overall, even increasing for several years in the early-2000s. Whilst the loch has perhaps the highest breeding density of Mallard in Scotland (Forrester & Andrews, 2007, p. 203), field surveys consistently suggest that there has not been any significant change in the size of the breeding population there since surveys were first conducted in the 1960s (Lauder, 2007). Thus, trends at Loch Leven, including the relative stability of wintering numbers in recent years compared with the clear decline at the Scottish level, are unlikely to be a result of changes in local breeding productivity. At the European scale (Burfield & van Bommel, 2004), the very large breeding population of Mallard (more than 3,300,000 pairs) was stable between 1970 and 1990. A decline was observed in some countries during 1990–2000, although most key European populations were stable and so the decline at the European scale will be small.

*Coot*

Whilst winter Coot numbers in GB have increased since the 1960s, perhaps dramatically so since 1994, those in Scotland are thought to be declining (Forrester & Andrews, 2007, p. 537). Indeed, data presented here show a consistent decline in Coot numbers in Scotland since the early-1990s. Some of this decline may be the result of range contractions and reduced breeding numbers, but the most likely cause is that fewer birds from northern Europe are wintering here, perhaps because of the recent mild winters (Forrester & Andrews, 2007, p. 357). At the European scale (Burfield & van Bommel, 2004), the very large breeding population of Coot (more than 1,300,000 pairs) increased markedly between 1970 and 1990 and, whilst remaining stable or increasing in some countries during 1990–2000, several key European populations declined. As such, the species underwent a moderate decline (over 10%) overall. At the local level, Coots are particularly sensitive to changes in macrophyte density (Perrow et al., 1997). Thus, whilst the continued increase in Coot numbers at Loch Leven during the winter cannot be a result of local breeding, because the number breeding there has declined sharply in recent years (Forrester & Andrews, 2007, pp. 534–535), it could, in part at least, be due to an 'improvement' in density of macrophyte at the loch (May & Carvalho, 2010). Indeed specific functional groups of waterfowl, such as herbivores including the Coot, have been reported to respond positively to high macrophyte cover and clear water in several shallow lakes elsewhere (e.g. see Hargeby et al., 1994; Hansson et al., 2010).

*Great Crested Grebe*

Total wintering numbers of Great Crested Grebe in Scotland were apparently stable between 1983/4 and 2002/3 (Forrester & Andrews, 2007, p. 349). However, the present study shows a consistent decline in average winter numbers of this bird since about 2004 (Fig. 2e). Whilst little is known of the origins of the Scottish winter population, local breeding birds are probably augmented by birds from England, Ireland or the nearby countries of continental Europe, depending on the severity of the weather and the availability of ice-free waters (Forrester & Andrews, 2007, p. 348). The decline in the Scottish wintering population could, thus, be a consequence of a recent series of milder winters (Carvalho & Kirika, 2003). At the local scale, declines in wintering populations in one part of Scotland appear to be balanced by increases elsewhere—but whether these are actually linked is unknown (Forrester & Andrews, 2007, p. 349). Thus, the maintenance of Loch Leven numbers, in recent years in the face of a Scottish level decline, might suggest its increased (or increasing) attractiveness as a wintering site. There is some evidence that this 'attractiveness' may be due to the presence of ice-free freshwater—with the recent run of mild winters precluding the need for birds to abandon the loch and move to the coast (see, for example, Forrester & Andrews, 2007, p. 347). At the European scale (Burfield & van Bommel, 2004), the large breeding population of Great Crested Grebe (more than 300,000 pairs) increased markedly between 1970 and 1990 and, whilst remaining stable or increasing across much of Europe during 1990–2000, certain European populations declined and so the species underwent a moderate decline (over 10%) overall.

*Tufted Duck*

The Scottish and local trends in average winter Tufted Duck numbers have diverged strikingly since the early-1990s (Fig. 2g). It seems likely that many of the breeding Tufted Duck in Scotland migrate to the south and west in winter. Thus, most of the Tufted Duck that are seen in Scotland in winter are thought to be from Iceland and Fennoscandia (Forrester & Andrews, 2007, p. 234). The decline in wintering numbers at the Scottish level probably relates to a reduction in the number of winter immigrants from these regions, perhaps caused by relatively mild winter weather. Similarly, the reasons for the continued winter increase in numbers at Loch Leven are unclear. The increase is unlikely to be related to breeding output at the loch, because breeding numbers there have declined over this period (Lauder, 2007). Winter food availability (perhaps linked with changes in macrophytes) and/or consistent, ice-free winters may have contributed to the increased wintering numbers at Loch Leven. At the European scale (Burfield & van Bommel, 2004), the large breeding population of Tufted Duck (more than 730,000 pairs) was stable between 1970 and 1990 and, whilst remaining stable or increasing across much of Europe during 1990–2000, there were declines in several north-eastern European countries. As such, the species underwent a moderate decline (over 10%) overall. At the European scale, this species is now provisionally evaluated as declining.

*Common Pochard*

Much of the Scottish decline in Pochard during the 1970s can be explained by the disappearance of a large flock from the Edinburgh area that used to feed around the city's sewerage outfalls in the Firth of Forth (Fox & Salmon, 1988). Improvements in sewage treatment also had knock-on effects on wintering numbers across the region, including at Loch Leven (Forrester & Andrews, 2007, p. 226). Whilst breeding in small numbers in Scotland, winter numbers of Pochard are boosted here by passage and winter visitors from Iceland and the northern continent (Forrester & Andrews, 2007, p. 224). Presumably, fewer of these birds are coming to Scotland in the winter, thus, contributing to the national decline in Pochard. Some of the earlier local increases in Pochard numbers at

Loch Leven were thought to be due to improved water quality there, especially reduced nutrient input and associated changes in the extent of macrophyte beds (May & Carvalho, 2010). The macrophyte *Chara* spp. is particularly important in the diet of Pochard throughout the palearctic, whilst another stonewort, *Nitella* spp., and various *Potamogeton* species (pond weeds) are also commonly eaten by this bird species (Cramp & Simmons, 1977, p. 565). The consistent increase in overwintering Pochard numbers at Loch Leven, in contrast to the Scottish-scale situation, could be a result of improved macrophyte conditions there. In addition, of the three diving ducks at Loch Leven, Pochard are known to be the most dependent upon macrophyte vegetation as a food source (Stewart & Lauder, 1997). At the European scale (Burfield & van Bommel, 2004), the large breeding population of Pochard (more than 210,000 pairs) was stable between 1970 and 1990 and, whilst remaining stable or increasing across much of Europe during 1990–2000, certain European populations, including key ones in Russia, declined. As such, the species underwent a moderate decline (over 10%) overall. At the European scale, this species is now provisionally evaluated as declining.

## General overview and future research directions

As can be seen from the above, the reasons for winter population changes—at either the local Loch Leven or national Scottish (or GB) scales—are seldom, if ever, clear for any of the waterfowl species examined. This is true whether or not long-term local and national trends are similar.

When local and national trends were similar, and Loch Leven was assumed to be merely 'following' a more geographically widespread trend, the likely causes were found to vary from the local to the national or international level. For example, changes in numbers were thought to be due to widespread climatic factors—especially the recent run of mild winters and associated low mortality and increase in juvenile survival rates (Tufted Duck, Mute Swan), reduced local shooting pressure in a variety of sites (Tufted Duck and possibly cormorant), changes in prey abundance or availability (Cormorant, pink-footed and greylag geese) and more widespread changes in the availability of food (pink-footed and greylag geese) that occurred as agricultural practices changed. It is likely that some of these factors are inter-linked (e.g. changes in food availability and wider climatic changes), suggesting that teasing apart key drivers is not a trivial task.

One of the aims of this study was to identify discontinuities in local versus national trends in waterfowl abundance, as a precursor to identifying the drivers of numerical changes at Loch Leven and maximising the effectiveness of management intervention, there. Interestingly, for dissimilar Scottish (or GB) versus Loch Leven trends, each of the five cases studied showed trends at the Scottish level that were declining in recent years while those at Loch Leven were stable (two cases: Mallard, Great Crested Grebe) or increasing (three cases: Coot, Tufted Duck, Pochard). However, like the similar trends described above, attributing likely causes to either local or national trends was difficult, as was explaining why there were local versus national divergence between them. Furthermore, in each of these five cases where wintering numbers were declining at the Scottish scale and those at Loch Leven were concurrently either stable or increasing, pan-European breeding populations showed overall declining trends (Burfield & van Bommel, 2004). Whilst the conservation status of three of these species was still considered to be 'secure' at the European scale, the status of both Tufted Duck and Pochard is now provisionally evaluated as 'declining' (Burfield & van Bommel, 2004). The Loch Leven wintering populations of both Tufted Duck and Pochard have continued to increase or have increased sharply, respectively, since the 1990s, perhaps suggesting that their relative conservation value at this site is increasing.

It is interesting to note that the same phenomenon (e.g. a run of mild winters) could reduce the number of birds wintering in Scotland (i.e. fewer continental immigrants), but increase those locally at Loch Leven (e.g. through the maintenance of ice-free freshwaters). Similarly, in some cases (e.g. Tufted Duck), wintering numbers at Loch Leven were not being maintained or increased, in part at least, through local breeding productivity at the loch because breeding numbers there seem to have declined markedly. In the case of Tufted Duck, this is due to predation on the main island colony. However, productivity has not been systematically monitored at Loch Leven and only in recent years has Scottish Natural Heritage changed bird monitoring methods to enable the extraction of comparable data and allow long-term changes to be tracked. In these cases, numbers of birds wintering at Loch Leven must be maintained through specific site-selection choices. In relation to such choices, in some cases Loch Leven might have got 'better' in recent years relative to other sites—for instance those with more abundant macrophytes—thus, becoming more attractive to waterfowl (e.g. Coot and Pochard). However, we cannot be certain that this is the case. An alternative (currently untested) explanation is merely that other sites are becoming 'worse', relative to Loch Leven. Whilst it is encouraging to think that management at Loch Leven is having positive effects on waterfowl numbers there, the relationship between the loch and the wider environment—be it Scotland, Great Britain or the wider European continent—is not clear in terms of supporting wintering waterfowl.

In the light of the difficulties identifying reasons for changes in populations of wildfowl at Loch Leven, management at the site is focused upon preventing long-term degradation and, where possible, optimising habitat quality. Catchment management to reduce inputs of nutrients, especially phosphorus, to the site and to improve water quality can be expected to benefit the ecosystem as a whole. These measures are implemented with a view to creating a site that is resilient to future change. Where future changes are driven by factors at the national, or even international, level, local scale management cannot address these issues.

Overall, for all five waterfowl species that have dissimilar local versus national trends, it is not clear how local changes in their numbers relate to changes in local feeding conditions or reflect changes in the abundance of migrating birds. For all of these species (cf. Forrester & Andrews, 2007, p. 226), a study of their feeding ecology is needed to determine how changes in local food resources interact with macro-scale population dynamics to influence local and regional patterns of abundance. Similarly, re-establishing ringing effort and analysing previously collected data would help elucidate bird movements within and to/from Scotland and provide information on winter survival rates. Marked populations would offer the opportunity to link annual changes in numbers and breeding success to factors such as rises in water temperature, and changes in water quality and vegetation. Whilst it may be possible to compare appropriate, relevant datasets for Loch Leven, there are clearly problems associated with interpreting such correlative type studies—particularly the difficulty in assessing the 'biological significance' of any observed correlations, as Savard et al. (1994) showed for waterfowl–wetland relationships in British Columbia. Thus, some form of experimental approach to investigate Loch Leven's bird-habitat relationships is also recommended. Overall, such research would address many gaps in our current knowledge of waterfowl ecology at Loch Leven. It would contribute to our theoretical understanding of the relative roles of local food supply, fluctuating environmental conditions and longer distance movements as drivers of waterfowl ecology, whilst also offering practical lessons for effective waterfowl management at this site and elsewhere.

**Acknowledgments** The authors would like to acknowledge Paul Brookes and Jeremy Squire (Scottish Natural Heritage) for continuing the bird counts at Loch Leven. We would also like to thank Willie Wilson, Jamie Montgomery and Sir David Montgomery for supporting this study. We would like to thank the British Trust for Ornithology for providing the national trend data and Alistair Dawson and Alan Lauder for useful comments that lead to the improvement of the manuscript. Finally, we would like to give credit to Alan Allison and Gordon Wright for maintaining the Loch Leven long-term bird monitoring programme in the "early years". This research was funded by the National Environment Research Council, UK and Scottish Natural Heritage. Loch Leven is part of the UK Environmental Change Network (http://www.ecn.ac.uk/).

## References

Allison, A. & I. Newton, 1974. Waterfowl at Loch Leven, Kinross. Proceedings of the Royal Society of Edinburgh B 74: 365–381.

Austin G.E., M. P. Collier & M. M. Rehfisch, 2008. 2007 Non-estuarine Coastal Waterbird Survey: population estimates and broad comparisons with previous surveys. British Trust for Ornithology (BTO) Research Report No. 501. BTO, Thetford.

Bailey-Watts, A. E. & A. Kirika, 1999. Poor water quality in Loch Leven (Scotland) in 1995in spite of reduced phosphorus loadings since 1985: the influences of catchment management and inter-annual weather variation. Hydrobiologia 403: 135–151.

Boere, G. C., C. A. Galbraith & D. A. Stroud (eds), 2006. Waterbirds Around the World. The Stationery Office, Edinburgh.

Bruun, B. & A. Singer, 1978. Guide to Birds of Britain and Europe. Hamlyn, London: 320 pp.

Bucke, D., G. D. Cawley, J. F. Craig, A. D. Pickering & L. G. Willough, 2006. Further studies of an epizootic of perch *Perca fluviatilis* L. of uncertain aetiology. Journal of Fish Diseases 2: 297–311.

Burfield, I. & F. van Bommel (eds), 2004. Birds in Europe: population Estimates, Trends and Conservation Status. Bird Life International Conservation Series No. 12, Cambridge.

Carss, D. N. & M. Marquiss, 1997. The diet of Cormorants *Phalacrocorax carbo* in Scottish freshwaters in relation to feeding habitats and fisheries. Ekologia Polska 45: 207–222.

Carss, D. N. & M. Marzano, 2005. Reducing the conflict between cormorants and fisheries on a pan-European Scale (REDCAFE), Summary and National Overviews. Report to EU, Commissioned by DG XIV Directorate-General for Fisheries: 374 pp.

Carss, D. N., R. M. Bevan, A. Bonetti, G. Cherubini, J. Davies, D. Doherty, A. El Hilli, M. J. Feltham, N. Grade, J. P. Granadeiro, D. Grémillet, J. Gromadzka, Y. N. R. A. Harari, T. Holden, T. Keller, G. Lariccia, R. Mantovani, T. M. McCarthy, M. Mellin, T. Menke, I. Mirowska-Ibron, W. Muller, P. Musil, T. Nazirides, W. Suter, J. F. G. Trauttmansdorff, S. Volponi & B. Wilson, 1997. Techniques for assessing cormorant diet and food intake: towards a consensus view. Proceedings IV European Conference on Cormorants. Supplemento alle Richerche di Biologia della Selvaggina XXVI: 197–230.

Carvalho, L. & A. Kirika, 2003. Long-term patterns of change in physical, chemical and biological aspects of water quality at Loch Leven. Final report to Scottish Natural Heritage, Centre for Ecology and Hydrology: 41 pp.

Carvalho, L., C. Ferguson, B. M. Spears, I. D. M. Gunn, H. Bennion, A. Kirika & L May, 2011. Water quality of Loch Leven: responses to enrichment, restoration and climate change. Hydrobiologia. doi:10.1007/s10750-011-0923-x.

Cramp, S. & K. E. L. Simmons, 1977. Handbook of the Birds of Europe, the Middle East and North Africa: The Birds of the Western Palearctic, Vol. 1: ostrich to Ducks. Oxford University Press, Oxford: 722 pp.

Cramp, S. & K. E. L. Simmons, 1979. The Birds of the Western Palearctic, Volume II: hawks to Bustards. Oxford University Press, Oxford: 695 pp.

Dudley, B., I. D. M. Gunn, L. Carvalho, I. Proctor, M. T. O'Hare, K. J. Murphy & A. Milligan, 2011. Changes in aquatic macrophyte communities in Loch Leven – evidence of recovery from eutrophication? Hydrobiologia. doi:10.1007/s10750-011-0924-9.

Favero, M. & P. H. Becker, 2006. Effects of the North Atlantic oscillation and El Niño southern oscillation on return rates, body mass and timing of migration of common terns *Sterna hirundo* breeding in Germany. In Boere, G. C., C. A. Galbraith & D. A. Stroud (eds), Waterbirds Around the World. The Stationery Office, Edinburgh.

Ferguson, C. A., L. Carvalho, E. M. Scott, A. W. Bowman & A. Kirika, 2008. Assessing ecological responses to environmental change using statistical models. Journal of Applied Ecology 45: 193–203.

Finlayson, M., J. Lowry, M. Grazia Bellio, S. Nou, R. Pidgeon, D. Walden, C. Humphrey & G. Fox, 2006. Biodiversity of the wetlands of the Kakadu region, northern Australia. Aquatic Sciences 68: 374–399.

Forrester, R. & I. Andrews, 2007. The Birds of Scotland. Scottish Ornithologists' Club: 1634 pp.

Fox, A. D. & D. G. Salmon, 1988. Changes in non-breeding distribution and habitat of Pochard *Aythya ferina* in Britain. Biological Conservation 46: 303–316.

Gunn, I. D. M., M. T. O'Hare, P. S. Maitland & L. May, 2011. Trends in Loch Leven invertebrate communities. Hydrobiologia. doi:10.1007/s10750-011-0926-7.

Hald-Mortensen, P., 1995. Danske Skarvers Fodevalg. 1992–1994. Skov-og Naturstyrelsen, Copenhagen.

Hansson, L.-A., A. Nicolle, C. Brönmark, A. Hargeby, Å. Lindström & G. Andersson, 2010. Waterfowl, macrophytes, and the clear water state of shallow lakes. Hydrobiologia 646: 101–109.

Hargeby, A., G. Andersson, I. Blindow & S. Johansson, 1994. Trophic web structure in a shallow eutrophic lake during a dominance shift from phytoplankton to submerged macrophytes. Hydrobiologia 279(280): 83–90.

Kirby, J. S., D. G. Salmon, G. L. Atkinson-Willes & P. A. Cranswick, 1995. Index numbers for waterbird populations III. Long-term trends in abundance of wintering wildfowl in Great Britain, 1966/67–1991/2. Journal of Applied Ecology 32: 536–551.

Lauder, A., 2007 Trends in the breeding duck (Anatidae) populations at Loch Leven, Perth & Kinross, 1966–2004. Scottish Birds (2007): 55–63.

May, L. & L. Carvalho, 2010. Maximum growing depth of macrophytes in Loch Leven, Scotland, United Kingdom, in relation to historical changes in estimated phosphorus loading. Hydrobiologia 646: 123–131.

May, L. & B. M. Spears, 2011. Managing ecosystem services at Loch Leven, Scotland, UK: actions, impacts and unintended consequences. Hydrobiologia. doi:10.1007/s10750-011-0931-x.

May, L., L. H. Defew, H. Bennion & A. Kirika, 2011. Historical changes (1905–2005) in external phosphorus loads to Loch Leven, Scotland, UK. Hydrobiologia. doi:10.1007/s10750-011-0922-y.

Owen, M., G. L. Atkinson Willes & D. Salmon, 1986. Wildfowl in Great Britain. Cambridge.

Perrow, M. R., J. Schutten, J. R. Howes, T. Holzer, F. J. Madgwick & A. J. D. Jowitt, 1997. Interactions between coot (*Fulica atra*) and submerged macrophytes: the role of birds in the restoration process. Hydrobiologia 342(343): 241–255.

Rehfisch, M. M., G. E. Austin, S. N. Freeman, M. J. S. Armitage & N. H. K. Burton, 2004. The possible impact of climate change on the future distributions and numbers of waders on Britain's non-estuarine coast. IBIS 146: S70–S81.

Savard, J.-P. L., W. S. Boyd & G. E. J. Smith, 1994. Waterfowl–wetland relationships in the Aspen Parkland of British Columbia: comparison of analytical methods. Hydrobiologia 279(280): 309–325.

Scottish Natural Heritage, 2007. The Reserve Plan for Loch Leven National Nature Reserve, 2008–2015.

Scottish Natural Heritage, 2009. Cormorants at Loch Leven: facts and figures [available on internet at http://www.snh.org.uk/pdfs/scottish/taysclack/cormorantff.pdf].

Spears, B. M. & I. D. Jones, 2010. The long-term (1979–2005) effects of the North Atlantic oscillation on wind-induced wave mixing in Loch Leven (Scotland). Hydrobiologia 646: 49–59.

Spears, B. M., L. Carvalho, R. Perkins, A. Kirika & D. M. Paterson, 2011. Long-term variation and regulation of internal loading in Loch Leven. Hydrobiologia. doi:10.1007/s10750-011-0921-z.

Stewart, B. & A. Lauder, 1997. Gizzard contents of Pochard, Tufted Duck and Goldeneye from Loch Leven, Kinross, in winter 1994–95. Scottish Birds 19: 28–35.

Thom, V., 1986. Birds of Scotland. Scottish Ornithologists' Club: 1600 pp.

Underhill, L. G., 1989. Indices for waterbird populations. BTO Research Report 52. British Trust for Ornithology, Thetford.

Underhill, L. G. & R. Prŷs-Jones, 1994. Index numbers for waterbird populations. I. Review and methodology. Journal of Applied Ecology 31: 463–480.

Van Eerden, E., H. Bos & L. van Hulst, 2007. In the mirror of a lake. Rijkswaterstaat centre for water management on behalf of regional directorate Ijsselmeergebied: 283 pp.

Winfield, I. J., C. E. Adams, J. D. Armstrong, R. Gardiner, A. Kirika, J. Montgomery, B. M. Spears, D. C. Stewart, J. E. Thorpe & W. Wilson, 2011. Changes in the fish community of Loch Leven: untangling anthropogenic pressures. Hydrobiologia. doi:10.1007/s10750-011-0925-8.

Hydrobiologia (2012) 681:105–116
DOI 10.1007/s10750-011-0930-y

LOCH LEVEN RESEARCH

# Modelling the response of phytoplankton in a shallow lake (Loch Leven, UK) to changes in lake retention time and water temperature

**J. A. Elliott · L. Defew**

Published online: 10 November 2011


**Abstract** The phytoplankton community of Loch Leven in 2005 was modelled and subjected to a combination of different flushing rates and water temperatures in order to assess the lake's sensitivity to these two climatic drivers. Whilst the simulated annual mean total chlorophyll *a* proved relatively insensitive to these changes, at the species level marked changes were recorded. Some species responded positively to increased temperature (e.g. *Aulacoseira*), some negatively (e.g. *Asterionella*), whilst others were negatively affected by increased flow (e.g. *Aphanocapsa*) and others enhanced (e.g. *Stephanodiscus*). However, this relationship with flow was season dependent with, for example, a simulated increase in summer inflows actually benefiting some species through increased nutrient supply, whereas an equivalent increase in flow in wetter seasons would have negatively affected those species (i.e. through flushing loss). Overall, the simulations showed that the range of species types simulated in the community was sufficient for one species to always benefit from the changing niches created by the multiple climatic drivers applied in this study. The level of exploitation by such a species was only constrained by the nutrient carrying capacity of the system, which led to the overall dampened response in the total chlorophyll *a* measure, both at the annual and season scale. Thus, whilst overall biomass showed relatively little reaction to the two climatic drivers tested, the phytoplankton community composition responded markedly.

**Keywords** Phytoplankton · Retention time · Climate change · Cyanobacteria · PROTECH

## Introduction

There has been increasing interest in recent years in the impacts that climate change has had, or could have, on lake ecology (e.g. DeStasio et al., 1996; Carvalho & Kirika, 2003; Winder & Schindler, 2004; Elliott et al., 2006; Elliott & May, 2008). Whilst much of this research has focused on the direct and indirect effects of increased water temperature, relatively few studies (e.g. Reynolds & Lund, 1988; Bailey-Watts et al., 1990; Jones & Elliott, 2007) have examined the impact of changes in rainfall and the way in which it affects lake phytoplankton ecology by altering the hydrological retention time. The latter effect could be particularly important in influencing water quality

Guest editors: L. May & B. M. Spears / Loch Leven: 40 years of scientific research

J. A. Elliott (✉)
Centre for Ecology, Library Avenue, Bailrigg, Lancashire LA1 4AP, UK
e-mail: alexe@ceh.ac.uk

L. Defew
Centre for Ecology & Hydrology, Bush Estate, Penicuik, Midlothian EH26 0QB, UK

because an increase in retention time can be associated with an increase in cyanobacteria, particularly in the summer months (Reynolds, 1993; Carvalho et al., 2008; Paerl & Huisman, 2008). This study investigated the impact of changes in both retention time and water temperature upon the phytoplankton in Loch Leven using a process-based computer model, PROTECH (Phytoplankton RespOnses To Environmental Change) (Reynolds et al., 2001).

Loch Leven was chosen for this study because it is a well-studied water body that has already shown signs of being influenced by climate change (Carvalho & Kirika, 2003). It is also an important generator of income to the local economy, with an estimated value of \$US 2.4M $ann^{-1}$. In the past, however, severe blooms of cyanobacteria in the lake have sometimes reduced its economic value, with one particular event costing an estimated \$US 1.2M in lost revenue and \$US 0.25M in increased water treatment costs to downstream industry (LLAMAG, 1993). So, it is important to understand the sensitivity of the phytoplankton population in the lake to the various aspects of changing climate. The lake and its phytoplankton community have already been successfully simulated before by PROTECH (Elliott & May, 2008) in a study that looked at the relative effects of changing water temperature and nutrient supply (both nitrogen and phosphorus) upon the lake as it was in 1995. However, the 1995 study only included one direct climate driver, i.e. temperature. In the present study, PROTECH has been used to explore the sensitivity of the phytoplankton community of Loch Leven to the combined climate drivers of changing water temperature and retention time, using a more recent (especially in terms of nutrient supply to the lake) starting state of 2005 for the baseline simulation.

## Methods

### Site description

Loch Leven is a large, shallow, eutrophic lake (surface area 13.3 $km^2$, mean depth 3.9 m, maximum depth 25.5 m), located in east central Scotland, UK (56° 12′ N, 3° 22′ W; altitude 107 m). It has a catchment area of 145 $km^2$ and an annual retention time of 140–180 days (Bailey-Watts & Kirika, 1999).

### *Data*

The driving and validation data for the simulations were taken from 2005. This year was chosen due to the availability of detailed nutrient and flow data available for most of the major inflows to the loch. These measurements included stream discharge values and associated soluble reactive phosphorus (SRP), nitrate-nitrogen and silica concentrations collected at 8-daily intervals. In-lake SRP, nitrate-nitrogen, silica and chlorophyll *a* concentrations, and water temperature had been measured every 14 days and phytoplankton species biovolume was recorded monthly. Daily meteorological data (cloud cover, air temperature, air humidity, wind speed) were available from a meteorological station at Leuchars Airfield, 35 km north-east of the lake. These wind speed values were corrected to local conditions by applying a 20% reduction, following the relationships derived by Smith (1973) from wind speeds measured at Loch Leven and at RAF Leuchars at the same time.

### *PROTECH model description*

The PROTECH model has been developed and tested on a wide range of lakes and reservoirs around the world over the last two decades (Elliott et al., 2000, 2005, 2007; Lewis et al., 2002; Elliott & Thackeray, 2004). The fundamental core of the biological component of PROTECH (Reynolds et al., 2001; Elliott & Thackeray, 2004) is the basic state variable equation that determines the daily change in the chlorophyll *a* concentration ($\Delta X/\Delta t$, mg $m^{-3}$ $d^{-1}$) of each algal species:

$$\Delta X/\Delta t = (r' - S - G - D).X \qquad (1)$$

where $r'$ is the growth rate defined as a proportional increase over 24 h, $S$ is the loss due to settling out of the water column, $G$ is the loss due to *Daphina* grazing (species > 50 μm are not grazed) and $D$ is the loss due to dilution. The growth rate ($r'$, $d^{-1}$) is further defined by:

$$r' = \min\{r'_{(\theta,\mathrm{I})}, r'_{\mathrm{P}}, r'_{\mathrm{N}}, r'_{\mathrm{Si}}\} \qquad (2)$$

where $r'_{(\theta,\mathrm{I})}$ is the growth rate due to temperature and daily photoperiod and $r'_{\mathrm{P}}$, $r'_{\mathrm{N}}$, $r'_{\mathrm{Si}}$ are the growth rates determined by phosphorus, nitrogen and silica concentrations. The $r'$ values are species dependent and relate to the morphology of the alga. Thus, for each species within the model, the initial starting value of $X$ mg chlorophyll $a$ m$^{-3}$ d$^{-1}$ (Eq. 1) is modified on a daily time-step to predict change in the chlorophyll $a$ concentration in the water column (see Reynolds et al., 2001, for details).

*Phytoplankton simulations*

Following examination of the phytoplankton count data to identify the most abundant and common species, the following phytoplankton types were selected for use in the model: *Cryptomonas*, *Anabaena*, *Gomphosphaeria*, *Aphanocapsa*, *Microcystis*, *Stephanodiscus*, *Asterionella* and *Aulacoseira*. The chlorophyll $a$ biomass measurements presented in this study for these simulated species and their total chlorophyll $a$ represent an integrated mean over the top 3 m of the water column, as this mimics the sampling method used to collect the observed data. These data were also used to calculate the percentage contribution of the cyanobacteria species in the simulated community. PROTECH does not include equations to simulate the release of SRP from sediment, which is an important factor in the nutrient budget of Loch Leven (Spears et al., 2007). Therefore, extra SRP was added to the water column from 1st June to 30th September. This amounted to an extra 1.7 mg m$^{-3}$ per 0.1 m PROTECH depth layer per day. Another site-specific calibration was also applied to this simulation by modifying the mixed depth function in PROTECH. As PROTECH models the deepest point in the lake, full mixing of the water column over-emphasises the importance of this point in lakes that are predominantly shallow, like Loch Leven. So, the depth of mixing was restricted to a maximum of 14.7 m from the surface, following the approximation used in Elliott et al. (2005).

Using the 2005 simulation of Loch Leven, key factors were altered to test the effects of changing water temperature and daily discharge. This was achieved by incrementally changing the temperature in the PROTECH depth layers by 1°C to give a range of water temperatures that ranged from 1°C cooler to 4°C warmer. At the same time, the daily discharge was multiplied by 0.5, 0.75, 1.0, 1.5 or 2.0 to give a range of discharge values. In addition, the original daily concentrations of nutrients associated with these discharges were corrected to maintain the nutrient supply at a 50:50 ratio of point to diffuse sources; it is believed that this represents the annual mean split of nutrient sources at this site (May et al., 2011). The consequence of using this ratio is that half of the nutrient load entering the lake is from point sources and independent of changes in inflow discharge whereas the other half of the load is from diffuse sources that change with discharge such that increased flow increases the nutrient load from this source. For each of these scenarios, the daily 3 m integrated chlorophyll $a$ data was used to calculate an annual mean value and quarterly seasonal means for Spring (March–May), Summer (June–August) and Autumn (September–November). Winter (December–January) values were also calculated, but these showed little changes and, therefore, were not considered further in this study.

## Results

### Comparison with observed data

In order to validate the model, its output was compared with the available phytoplankton observations (Fig. 1). Total chlorophyll $a$ was often measured in three places in Loch Leven during the fortnightly sampling cycle, thus an estimate of variation in chlorophyll $a$ could be calculated as well as the mean value. These mean values were compared with the PROTECH output (Fig. 1a) and found to show a statistically significant fit ($R^2 = 0.42$, $P < 0.001$),

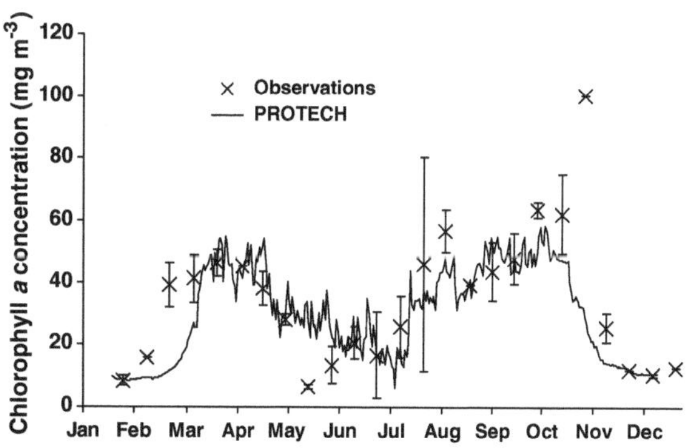


**Fig. 1** Comparison between observed (*crosses*) and PROTECH-simulated (*solid line*) annual mean total chlorophyll $a$ concentrations (mg m$^{-3}$) for Loch Leven, 2005. *Error bars* for observations, which are taken from multiple sites, represent 2 standard deviations (c. 95% confidence interval)

**Fig. 2** Comparison between observed (*crosses*) total biovolumes ($\mu m^3$ $ml^{-1}$) and PROTECH-simulated (*solid line*) chlorophyll *a* concentrations (mg $m^{-3}$) for the major taxa in Loch Leven in 2005: **a** diatoms; **b** cyanobacteria

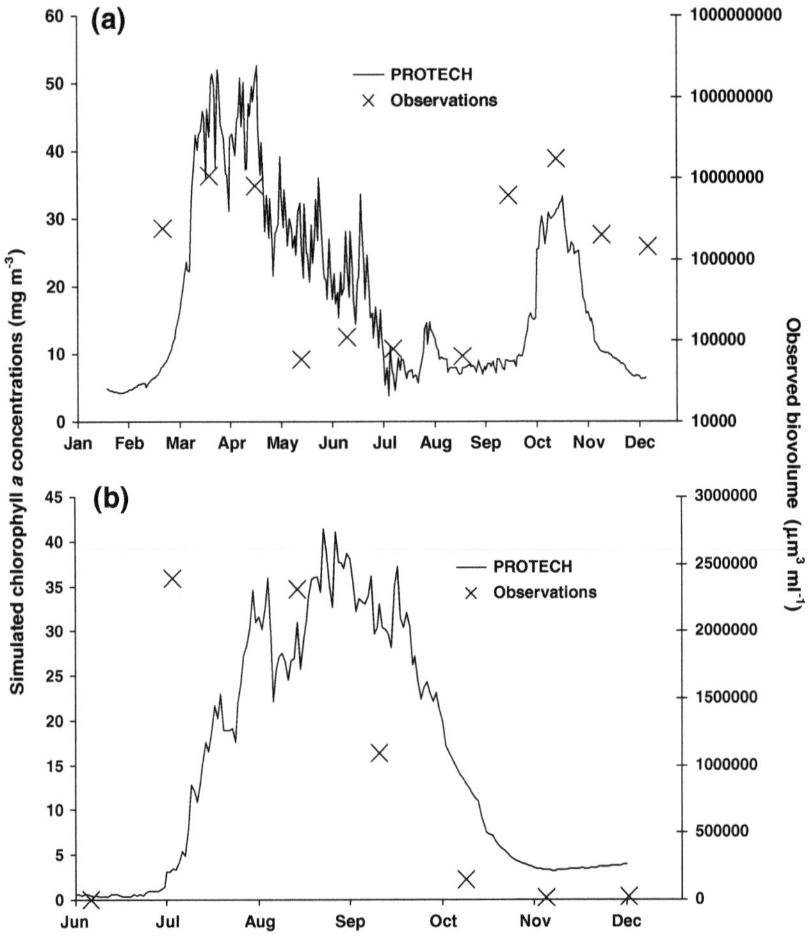


which improved further with the exclusion of the single value of 100.3 mg $m^{-3}$ recorded on 2nd November 2005 ($R^2 = 0.67$, $P < 0.001$). However, whilst the bimodal bloom pattern was recreated well, there were some notable differences. For example, the simulated spring bloom was about 2 weeks later and did not decline as sharply as the observed values in May.

The main simulated taxa were compared with the observed data (Fig. 2) and, because two different methods of enumeration were used (i.e. biovolume for observed data, chlorophyll *a* for PROTECH), only simple visual comparisons were made between the patterns. The model simulation captured the bimodal pattern of diatom development well throughout the year (Fig. 2a) and was broadly in agreement with the bloom of cyanobacteria in the second half of the year (Fig. 2b). However, the latter comparison did show that PROTECH's simulated changes in cyanobacteria biomass were consistently 2–3 weeks later in the year than those of the observed values. Nevertheless, this simulation was considered adequate for use as a baseline for scenario testing, particularly because the use of annual and seasonal means in the analysis would be less affected by the differences in timing highlighted above.

### *Changes in annual means of total and species chlorophyll a concentrations*

Predicted annual mean total chlorophyll *a* concentrations ranged between 24 and 31 mg $m^{-3}$ (Fig. 3a), changing relatively little with different flows and temperatures, although there was a slight tendency for the model to produce less biomass with the imposition of more extreme flow scenarios. However, such low levels of response in these annual means did not mean the community composition was unaffected.

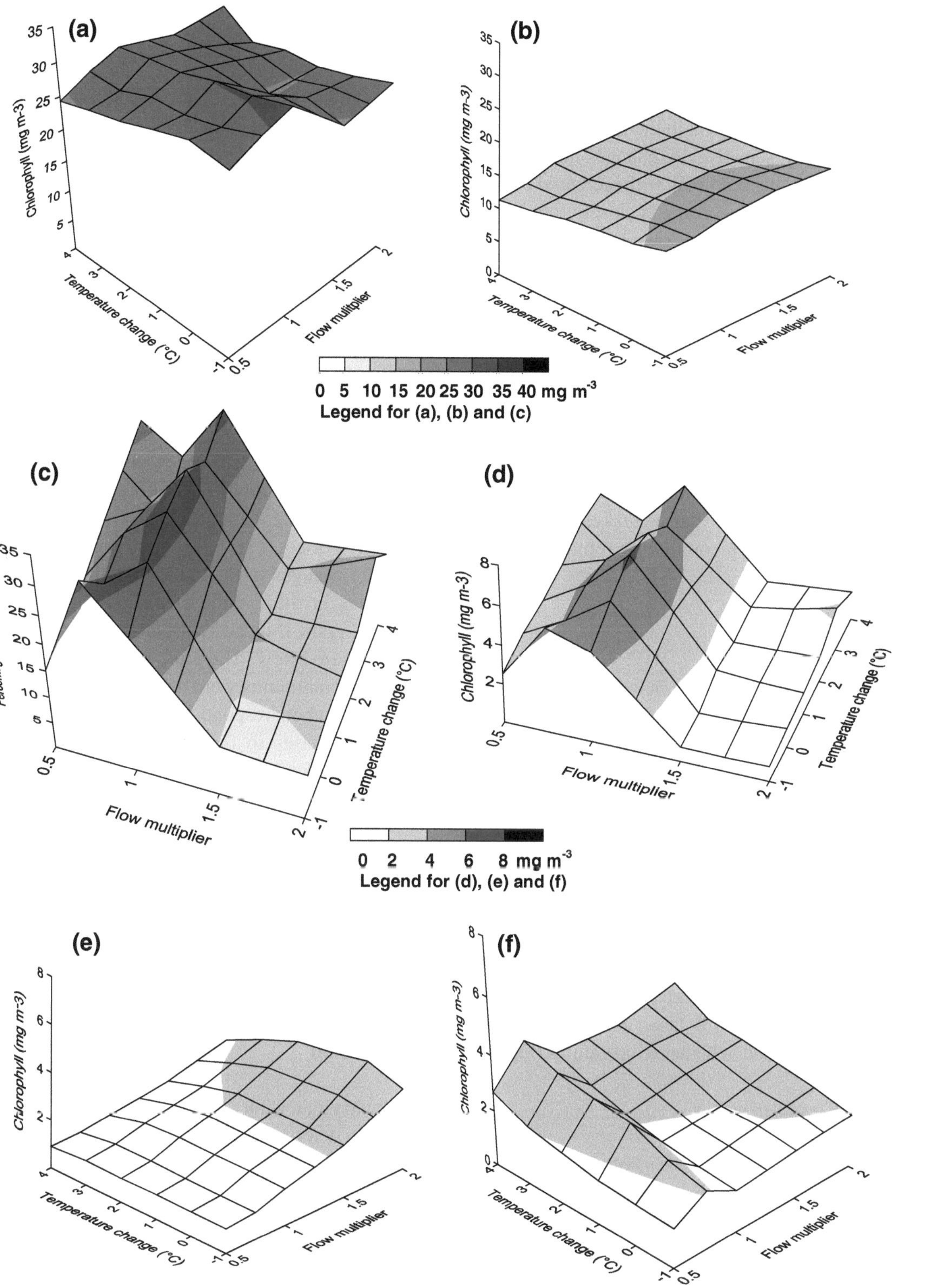


**Fig. 3** Impact of changing water temperature (°C) and discharge on the annual mean chlorophyll *a* concentration (mg $m^{-3}$) in Loch Leven for: **a** total chlorophyll; **b** *Asterionella*; **c** percentage cyanobacteria; **d** *Aphanocapsa*; **e** *Stephanodiscus*; **f** *Aulacoseira*. *Grey scale* legends denote chlorophyll *a* concentration (mg $m^{-3}$) bands

The simulated phytoplankton that equated to *Asterionella* produced most of the total chlorophyll *a* (Fig. 3b) and responded to increasing temperature by producing slightly less biomass over the year. The percentage of cyanobacteria in the community proved to be especially sensitive to changes in discharge (Fig. 3c), decreasing markedly with increased flushing rate. This pattern of response was closely followed by the cyanobacteria species, *Aphanocapsa* (Fig. 3d). The annual means of two other species are also worthy to mention because their responses were different from those discussed above. First, the small diatom, *Stephanodiscus*, produced more biomass with increasing flow (Fig. 3e). Second, the filamentous diatom, *Aulacoseira*, responded positively to increasing temperature (Fig. 3f), which was the opposite response to that of its closest competitor in the simulation, *Asterionella*. Overall, it was clear that, when the biomass produced over the whole year was taken into account, the lake phytoplankton community remained diatom dominated.

*Changes in spring means*

Dominance by diatoms in the simulations was even more marked in the spring period (March–May), when mean total chlorophyll *a* was at its highest (30–39 mg $m^{-3}$) and relatively unaffected by the scenarios (Fig. 4a). *Asterionella* was the dominant diatom, producing nearly all of the spring biomass (Fig. 4b), but this species showed a slight decline (c. 1 mg $m^{-3}$ per level of change) in mean chlorophyll *a* concentration with increasing flow. Two other diatoms made up the rest of the total biomass, with *Aulacoseira* showing an increase in mean spring biomass at high temperatures (Fig. 4c) and flows. Similarly, *Stephanodiscus* biomass increased rapidly with larger flows, although this effect was diminished at higher temperatures (Fig. 4d).

*Changes in summer means*

Mean total chlorophyll *a* in the summer period (June–August), was sensitive to both changes in temperature and flow regime, with ca. 20 mg $m^{-3}$ difference between the highest and lowest mean value (Fig. 5a). Most of this change was caused by increased flow reducing the mean chlorophyll *a* level.

In contrast to the pattern of change in total chlorophyll *a* concentration, *Asterionella* contributed far less to the overall biomass in this season than in spring and responded positively to increasing flow and decreasing temperature (Fig. 5b). This was because the total chlorophyll *a* comprised mainly cyanobacteria (Fig. 5c), especially *Aphanocapsa* (Fig. 5d). This species showed a marked sensitivity, decreasing in mean biomass with high and, to a lesser extent, low flows. Interestingly, it also showed a slight decline in abundance with increasing temperature. Finally, the flagellate *Cryptomonas* responded positively to decreasing flow and increased temperatures (Fig. 5e).

*Changes in autumn means*

The autumn (September–November) mean total chlorophyll *a* value was the second highest of the four seasons and its pattern of response was similar to that observed in the summer. The exception was that, under high water temperatures, flow ceased to have a negative effect (Fig. 6a). Again, the underlying cause for the changes in total chlorophyll *a* was due to the changes in the cyanobacteria community (Fig. 6b), which, again, showed a considerable negative response to increased flow. The most abundant diatoms in the autumn community were *Asterionella* and *Aulacoseira* but, whilst both responded little to the changes in flow, their responses to the changes in temperature contrasted markedly (Fig. 6c, d). *Asterionella* declined in biomass with increasing temperature, whilst *Aulacoseira* increased. Again, the dominant cyanobacterium was *Aphanocapsa,* which responded negatively to extreme flows (Fig. 6e). Finally, *Cryptomonas* produced a very unusual response surface (Fig. 6f) in which it responded positively when flow and temperature were both low and also when both were high.

## Discussion

Annual mean values of measured parameters are often used to determine the overall quality of a water body. In this context, measures of phytoplankton are often no more detailed than mean annual measures of total chlorophyll *a*. This study illustrates why such measures can sometimes be too simplistic. The results show that considerable changes in species composition can occur in response to changing temperature and flow scenarios, whilst the overall annual mean total chlorophyll *a* concentrations change very little

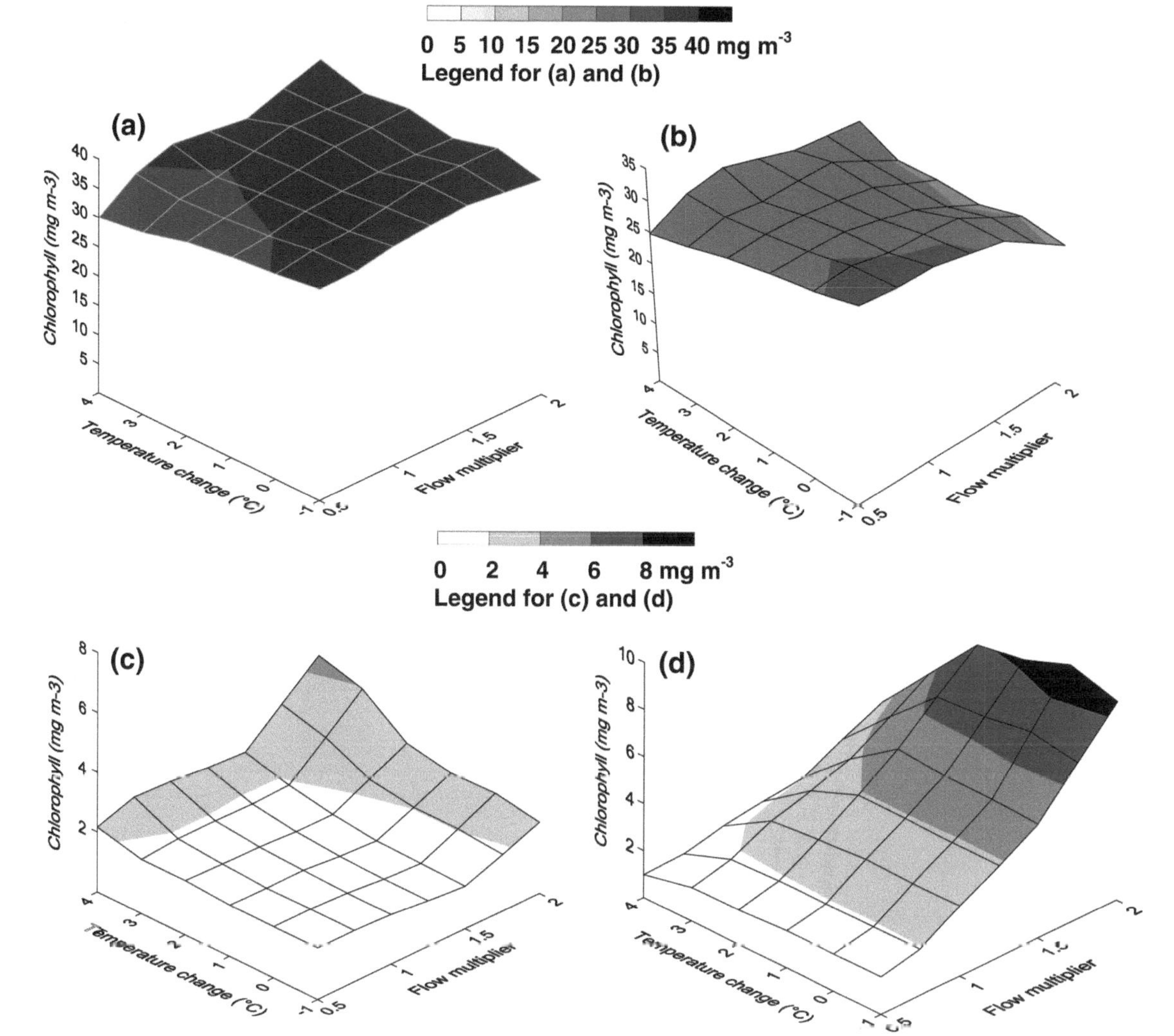


**Fig. 4** Impact of changing water temperature (°C) and discharge of the spring mean chlorophyll *a* concentration (mg m$^{-3}$) in Loch Leven for: **a** total chlorophyll *a*; **b** *Asterionella*; **c** *Aulacoseira*; **d** *Stephanodiscus*. *Grey scale* legends denote chlorophyll *a* concentration (mg m$^{-3}$) bands

(Fig. 3a). It was only when low temperatures (causing lower growth rates) were coupled with increased flushing losses of biomass and nutrients (caused by high flows) that total chlorophyll *a* showed a slight decline; the latter is supported by similar observations in other lakes and reservoirs (Kalff, 2002).

At the species level, a great deal of variation occurred in response to changes in temperature and flushing rate. Some species responded positively to increased temperature (e.g. *Aulacoseira*, Fig. 3f) and others negatively (e.g. *Asterionella*, Fig. 3b). Other species produced more biomass with higher flushing, such as the small, fast growing diatom *Stephanodiscus* (Fig. 3e), whilst slower growing species like *Aphanocapsa* were outcompeted under such conditions (Fig. 3d). These results showed that, as one species reacted negatively to the change in conditions, another species responded positively. This type of response pattern dampened the effect of the changes in temperature and flow on the total chlorophyll *a* concentration in the lake. Such species changes were even more evident when the seasonal trends were examined.

The spring period was dominated by diatoms and, again, the three main species responded differently to the environmental change scenarios. High flows negatively affected *Asterionella* but positively benefited

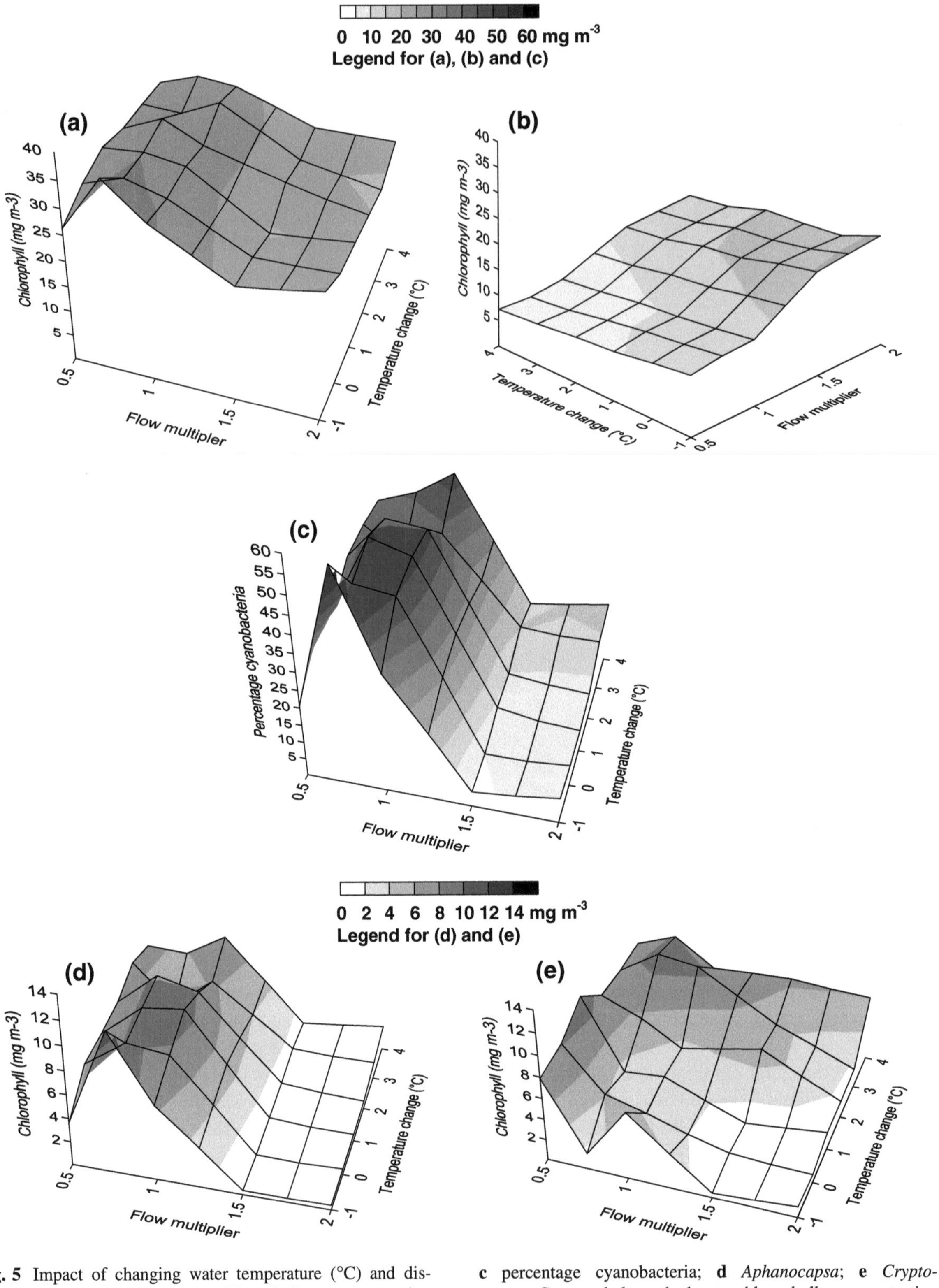


**Fig. 5** Impact of changing water temperature (°C) and discharge of the summer mean chlorophyll *a* concentration (mg $m^{-3}$) in Loch Leven for **a** total chlorophyll *a*; **b** *Asterionella*; **c** percentage cyanobacteria; **d** *Aphanocapsa*; **e** *Cryptomonas*. *Grey scale* legends denote chlorophyll *a* concentration (mg $m^{-3}$) bands

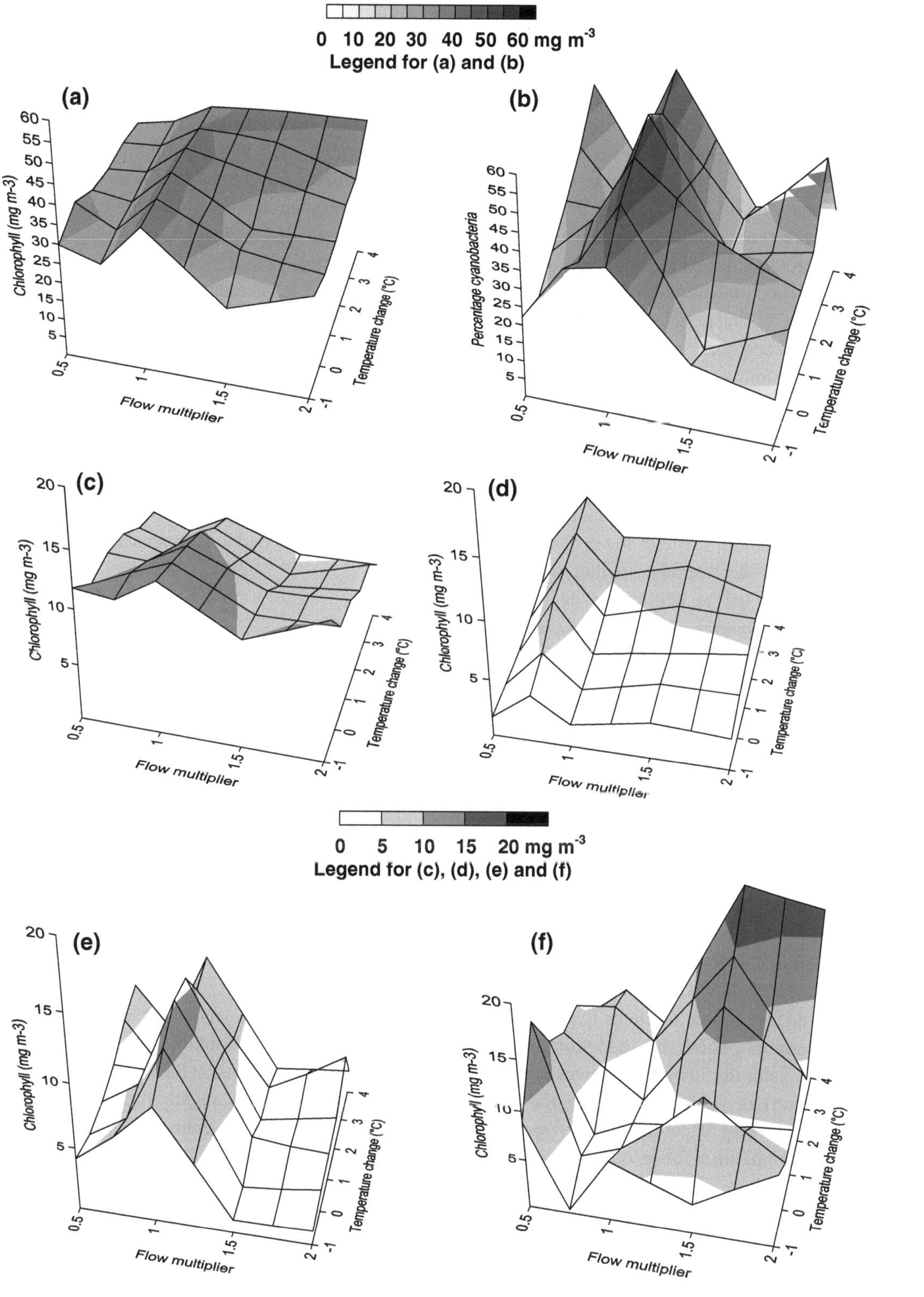


**Fig. 6** Impact of changing water temperature (°C) and discharge of the autumn mean chlorophyll *a* concentration (mg m$^{-3}$) in Loch Leven for: **a** total chlorophyll *a*; **b** percentage cyanobacteria; **c** *Asterionella*; **d** *Aulacoseira*; **e** *Aphanocapsa*; **f** *Cryptomonas*. *Grey scale* legends denote chlorophyll *a* concentration (mg m$^{-3}$) bands

*Stephanodiscus* and, to a much lesser extent, *Aulacoseira* (Fig. 4). It seems likely that this was due, once more, to the flushing losses caused by the increased flow that the relatively small sized, fast growing, *Stephanodiscus* was particularly suited to survive (provided other factors were not limiting its growth). Such a reaction whereby smaller phytoplankton prevail under conditions of high flushing is well documented (Boucher et al., 1984; Rojo et al., 1994; Kalff, 2002) and in previous PROTECH studies, where this phytoplankton functional type (i.e. relatively small species sensu the C functional group after Reynolds, 1995) has often been seen to increase in prevalence under such conditions (Jones & Elliott, 2007; Bernhardt et al., 2008).

The importance of seasonal means was highlighted further when the response of *Asterionella* in the summer was considered. At that time of year, its biomass was positively enhanced by increased flows, particularly when combined with lower temperatures (Fig. 5). This opposite response to that seen in the spring was related to the decline of its two main competitors (*Cryptomonas* and *Aphanocapsa*) under these conditions, as well as receiving a net benefit of increased nutrient supply via the larger inflows; this latter effect has been observed before for diatoms in Loch Leven, which benefited from additional nutrients entering during high flows (Bailey-Watts et al., 1990). Again, we see one species benefiting over another when more of the resources essential for growth (e.g. nutrients, light) become available. However, it was also clear that low flow scenarios could be relatively unfavourable for growth, with a decline in the two most dominant phytoplankton species (*Asterionella* and *Aphanocapsa*) being evident. This occurred because the reduced flow resulted in a reduction in the supply of diffuse source nutrients to the modelled lake. By the summer period, this reduction in supply, combined with naturally lower flows, was starting to become an important factor in reducing the overall phytoplankton carrying capacity and, subsequently, increasing the competition for resources.

With natural inflow discharges increasing in the autumn, the nutrient restriction was slightly alleviated. However, it remained an influential factor at low flows for most species (Fig. 6), whilst high flows still generally caused a decline in biomass. Intriguingly, temperature also started to have a more obvious effect on some of the mean chlorophyll *a* values. During the autumn period, increasing temperature reduced the biomass of *Asterionella* and increased that of *Aulacoseira*. These two species are close competitors of each other, both being well adapted to low light level conditions (R functional group sensu Reynolds, 1995) and, through most of the simulations, *Asterionella* was the dominant of the two. However, in this autumn period, higher temperatures increased the relative growth rate of *Aulacoseira* allowing it to produce more biomass than *Asterionella*. Finally, *Cryptomonas* produced a very diverse response in the autumn, with peaks in biomass occurring at high temperatures and flows and also under completely opposite conditions. *Cryptomonas* is a relatively large phytoplankter and, in PROTECH, is given specific motility characteristics to simulate its excellent nutrient scavenging abilities (Reynolds, 1984). Thus, it can thrive under low nutrient conditions, but needs elevated temperatures under high flow conditions to achieve the high growth rates needed to compensate for flushing losses.

## Conclusion

Temperature had little effect on the overall biomass produced by the phytoplankton in Loch Leven. This is in concurrence with the results of previous studies on the loch (Carvalho & Kirika, 2003; Elliott & May, 2008) and other experimental investigations (McKee et al., 2003; Moss et al., 2003). Furthermore, there was no evidence, either annually or seasonally, of temperature enhancing cyanobacteria abundance. Again, this was in agreement with the results of Elliott & May (2008), but is in contrast to those from a similar study where PROTECH was applied to Bassenthwaite Lake (Elliott et al., 2006).

Changes in flow and flushing rate had a greater effect on the phytoplankton community than changes in temperature, at least at the seasonal scale. When the rate of flow was changed, there appeared to be a general response of reduced biomass with extreme flows. In this instance, high flows acted by causing high flushing losses and low flows reduced the nutrient supply, particularly in the summer. However, this study also illustrates how some species possess traits that could enable them to exploit these conditions. Small, fast growing species, such as *Stephanodiscus* can recoup their flushing losses under high flow conditions. In contrast, other species, such as

*Aphanocapsa*, are ideally adapted to surviving under low nutrient conditions. Thus, as one species becomes stressed by a physical change (e.g. temperature, retention time), another species becomes better able to exploit the resources available. Such allogenic succession is readily recognised across ecosystems (Begon et al., 1996).

**Acknowledgements** The authors thank Sir David Montgomery for providing access to the loch and the authors are grateful to Kinross Estates, Scottish Natural Heritage and Scottish Environment Protection Agency staff for their help with fieldwork. The authors thank the British Atmospheric Data Centre for providing the meteorological data and Stephen Thackeray for help with drawing the figures. The authors also thank Linda May and Laurence Carvalho for their constructive comments and the latter for providing the phytoplankton biovolume data for 2005. This research was funded by Scottish Natural Heritage, the Scottish Environment Protection Agency and the Natural Environment Research Council, UK. Loch Leven is part of the UK Environmental Change Network (http://www.ecn.ac.uk/).

## References

Bailey-Watts, A. E. & A. Kirika, 1999. Poor water quality in Loch Leven (Scotland) in 1995 in spite of reduced phosphorous loadings since 1985: the influences of catchment management and inter-annual weather variation. Hydrobiologia 403: 135–151.

Bailey-Watts, A. E., A. Kirika, L. May & D. H. Jones, 1990. Changes in phytoplankton over various time scales in a shallow eutrophic lake: the Loch Leven experience with special reference to the influence of flushing rate. Freshwater Biology 23: 85–111.

Begon, M., J. L. Harper & C. R. Townsend, 1996. Ecology: Individuals, Populations and Communities, 3rd ed. Blackwell Scientific Publications, Oxford.

Bernhardt, J., J. A. Elliott & I. D. Jones, 2008. Modelling the effects on phytoplankton communities of changing mixed depth and background extinction coefficient on three contrasting lakes in the English Lake District. Freshwater Biology 53: 2573–2586.

Boucher, P., D. W. Blinn & D. B. Johnson, 1984. Phytoplankton ecology in an unusually stable environment (Montezuma Well, Arizona, U.S.A.). Hydrobiologia 119: 149 160.

Carvalho, L. & A. Kirika, 2003. Changes in shallow lake functioning: response to climate change and nutrient reduction. Hydrobiologia 506: 789–796.

Carvalho, L., B. Spears, B. Dudley, I. Gunn, S. Zimmermann, L. Defew & L. May, 2008. Loch Leven 2007: trends in water quality and biological communities. Centre for Ecology & Hydrology: 24 pp.

DeStasio B.T.Jr., D. K. Hill, J. M. Kleinhans, N. P. Nibbelink & J. L. Magnasun, 1996. Potential effects of global climate change on small north-temperate lakes: physics, fish and plankton. Limnology and Oceanography 41: 1136–1149.

Elliott, J. A. & L. May, 2008. The sensitivity of phytoplankton in Loch Leven (UK) to changes in nutrient load and water temperature. Freshwater Biology 53: 32–41.

Elliott, J. A. & S. J. Thackeray, 2004. The simulation of phytoplankton in shallow and deep lakes using PROTECH. Ecological Modelling 178: 357–369.

Elliott, J. A., A. E. Irish, C. S. Reynolds & P. Tett, 2000. Modelling freshwater phytoplankton communities; an exercise in validation. Ecological Modelling 128: 19–26.

Elliott, J. A., C. Escot, A. Basanta-Alves & L. Cruz-Pizarro, 2005a. Simulations of phytoplankton dynamics in El Gergal reservoir, southern Spain (PROTECH). In Rouen K. & G. George (eds), New Tools for the Monitoring, Modelling & Management of Fresh Waters. Freshwater Forum 23: 78–92.

Elliott, J. A., S. J. Thackeray, C. Huntingford & R. G. Jones, 2005b. Combining a Regional Climate Model with a phytoplankton community model to predict future changes in phytoplankton in lakes. Freshwater Biology 50: 1404–1411.

Elliott, J. A., I. D. Jones & S. J. Thackeray, 2006. Testing the sensitivity of phytoplankton communities to changes in water temperature and nutrient load, in a temperate lake. Hydrobiologia 559: 401–411.

Elliott, J. A., I. Persson, S. J. Thackeray & T. Blenckner, 2007. Phytoplankton modelling of Lake Erken, Sweden by linking the models PROBE and PROTECH. Ecological Modelling 202: 421–426.

Jones, I. D. & J. A. Elliott, 2007. Modelling the effects of changing retention time on abundance and composition of phytoplankton species in a small lake. Freshwater Biology 52: 988–997.

Kalff, J., 2002. Limnology: Inland Water Ecosystems. Prentice Hall, Upper Saddle River.

Lewis, D. M., J. A. Elliott, M. F. Lambert & C. S. Reynolds, 2002. The simulation of an Australian reservoir using a phytoplankton community model (PROTECH). Ecological Modelling 150: 107–116.

LLAMAG, 1993. Report of Loch Leven Area Management Advisory Group. Forth River Purification Board, Edinburgh.

May, L., L. H. Defew, H. Bennion & A. Kirika, 2011. Historical changes (1905–2005) in external phosphorus loads to Loch Leven, Scotland, UK. Hydrobiologia. doi: 10.1007/s10750-011-0922-y.

McKee, D., D. Atkinson, S. E. Collings, J. W. Eaton, A. B. Gill, I. Harvey, K. Hatton, T. Heyes, D. Wilson & B. Moss, 2003. Response of freshwater microcosm communities to nutrients, fish, and elevated temperature during winter and summer. Limnology & Oceanography 48: 707–722.

Moss, B., D. McKee, D. Atkinson, S. E. Collings, J. W. Eaton, A. B. Gill, I. Harvey, K. Hatton, T. Heyes & D. Wilson, 2003. How important is climate? Effects of warming, nutrient addition and fish on phytoplankton in shallow lake microcosms. Journal of Applied Ecology 40: 782–792.

Paerl, H. W. & J. Huisman, 2008. Blooms like it hot. Science 320: 57–58.

Reynolds, C. S., 1984. The Ecology of Freshwater Phytoplankton. Cambridge University Press, Cambridge.

Reynolds, C. S., 1993. Scales of disturbance and their roles in plankton ecology. Hydrobiologia 249: 151–171.

Reynolds, C. S., 1995. Successional change in the planktonic vegetation: species, structures, scales. In Joint, I. (ed.), The Molecular Ecology of Aquatic Microbes. Springer, Berlin: 115–132.
Reynolds, C. S. & J. W. G. Lund, 1988. The phytoplankton of an enriched, soft-water lake subject to intermittent hydraulic flushing (Grasmere, English Lake District). Freshwater Biology 19: 370–404.
Reynolds, C. S., A. E. Irish & J. A. Elliott, 2001. The ecological basis for simulating phytoplankton responses to environmental change (PROTECH). Ecological Modelling 140: 271–291.
Rojo, C., M. Alvarez Cobelas & M. Arauzo, 1994. An elementary, structural analysis of river phytoplankton. Hydrobiologia 289: 43–55.
Smith, I. R., 1973. The assessment of winds at Loch Leven, Kinross. Weather 28: 202–210.
Spears, B. M., L. Carvalho & D. M. Paterson, 2007. Phosphorus partitioning in a shallow lake: implications for water quality management. Water and Environment Journal 21: 27–53.
Winder, M. & D. E. Schindler, 2004. Climatic effects on the phenology of lake processes. Global Change Biology 10: 1844–1856.

Hydrobiologia (2012) 681:117–130
DOI 10.1007/s10750-011-0931-x



Opinion Paper

# Managing ecosystem services at Loch Leven, Scotland, UK: actions, impacts and unintended consequences

L. May · B. M. Spears

Published online: 9 November 2011


**Abstract** Rivers, lakes and wetlands are good examples of ecosystems that provide multiple, concurrent, services to mankind. Human society has often exploited these systems by enhancing one ecosystem service at the expense of another. Loch Leven, Scotland, UK, is a good example of this. Over the past 150 years, the lake has been subjected to hydrological modification, fish stocking and pollution control to improve the delivery of key goods and services. This study uses historical records to explore the results of these interventions on the ecosystem services that were targeted for improvement and the knock-on effects on other services provided by the lake. The results suggest that, when management changes are being considered to enhance particular ecosystem services, the potentially damaging effects on other ecosystem services should be taken into account. This requires a better understanding of the role of ecosystem function in delivering ecosystem services, and of the links between multiple ecosystem services, than is currently available. While further research is clearly needed, the value of long-term datasets in providing knowledge and understanding through 'hindsight' should not be underestimated. The study concludes that successful management actions are likely to be those that incorporate lessons learned from previous decisions.

**Keywords** Lake management · Brown trout · Rainbow trout · *Daphnia* · Nutrient load · Pesticide pollution · Cost

## Introduction

Lakes, rivers and wetlands are good examples of ecosystems that provide a wide range of ecosystem services at the same time. Ecosystem services are the benefits that people obtain from the natural environment (MEA, 2005). They include provisioning services (e.g. delivery of food, water, pharmaceuticals, energy), regulating services (e.g. carbon sequestration, climate regulation, water purification, disease control), supporting services (e.g. nutrient cycling, seed dispersal, primary production) and cultural services (e.g. providing inspiration, facilitating recreation, enabling scientific discovery). These services are produced by complex processes and interactions that are intimately linked in such a way that the exploitation of one ecosystem service can cause knock-on effects on many others (Heal et al., 2001; Pereira et al., 2005; Reid et al., 2006).

In the past, human society has often exploited these systems by enhancing one ecosystem service at the

Guest editors: L. May & B. M. Spears / Loch Leven: 40 years of scientific research

L. May (✉) · B. M. Spears
Centre for Ecology and Hydrology, Penicuik, Midlothian EH26 0QB, UK
e-mail: lmay@ceh.ac.uk

expense of another. For example, water has been used for consumption, irrigation or transport purposes, with little consideration of the impact that this may have on other services, such as food supply, flood control, purification of human and industrial wastes, and provision of habitat to support plant and animal life (Baron et al., 2002). This has led to the degradation or loss of many of the services that mankind depends on, including those that are difficult, if not impossible, to replace (Huisman, 1995). This is an important issue because the degradation or loss of any part of an ecosystem can reduce its resilience, i.e. its capacity to adapt to environmental alterations, such as climate change (Folke et al., 2004).

Loch Leven, a shallow eutrophic lake in the lowlands of Scotland, UK, is a good example of a waterbody that supplies a wide range of ecosystem services across local, national and international scales. Over the past 150 years, this system has been subjected to a range of management initiatives that have sought to improve some of the key goods and services provided by the lake. These activities have been carefully documented in terms of their original aims and subsequent outcomes, generating a wealth of historical information that enables us to explore the effects of these management activities with the benefit of hindsight. This study uses these records to investigate, not only the results of management intervention on the ecosystem service that was targeted for improvement, but also the knock-on effects on other services that the lake provides. These results are reviewed in terms of their wider implications for the sustainable management of other complex systems and the ecosystem services that they deliver.

## Site description

Loch Leven is a shallow, eutrophic lake lying at about 107 m above sea level in the lowlands of Scotland, UK (56°10′N; 3°30′W). It has a surface area of 13.3 $km^2$, and mean and maximum depths of 3.9 and 25.5 m, respectively (Kirby, 1971). The lake is a world famous trout fishery, internationally recognised conservation area (SSSI, Ramsar, SAC, Natura 2000) and a source of water supply to downstream industry (May & Spears, 2011).

The lake lies within a catchment that covers an area of about 145 $km^2$ and ranges in altitude from 107 to 482 m.a.o.d. Most of the catchment is intensively farmed, with the best quality land being used for high-value potato and vegetable crops, as well as cereal and oilseed rape (Castle et al., 1999). Livestock rearing (cattle and sheep) is mainly restricted to the more upland areas, which tend to be further away from the lake (LLCMP, 1999). However, there is a small area of intensive poultry production near to one of the main inflows, the South Queich. A small amount (ca. 11%) of the catchment is wooded and the remainder (ca. 2%) is used for habitation.

The catchment is sparsely populated, with only about 11,000 people living in the area (Frost, 1996). Of these, about 60 % live in the towns of Kinross, Milnathort and Kinnesswood (Perth & Kinross Council, 2004), in properties that are served by mains sewerage systems. A further 650 households, in the more remote parts of the catchment, rely upon septic tank systems for the management of their domestic waste (Dudley & May, 2007).

In general, there is little industry within the catchment. However, there have been woollen mills on the banks of one of the main inflows to the lake, the South Queich, since 1840 (Munro, 1994). Until recently, at least one of these mills discharged large quantities of industrial effluent into the lake (D'Arcy et al., 2006). In addition, sand and gravel extraction is practiced in some areas of the catchment. This is believed to have adversely affected some of the traditional trout spawning beds and nursery areas along the inflows to the lake (Montgomery, 1994).

## Management activities to enhance key ecosystem services

Like all lakes, Loch Leven is the 'sink' into which the upstream catchment drains (Baron et al., 2002). As such, both the lake and its catchment are inextricably linked in terms of the wide range of ecosystem services that they provide and the complex nature of their interactions. This study focuses on the key ecosystem services that the lake provides, which are:

Provisioning services

- food (fish)
- water supply to downstream industry

Regulating services

- waste management
- water purification

- flood control

Supporting services

- nutrient cycling
- primary production

Cultural services

- cultural, intellectual and spiritual inspiration
- recreation, tourism and nature conservation
- scientific discovery

Many attempts have been made to enhance some of these services over the last 150 years through targeted management activities. These have focused, primarily, on improving the water supply to downstream industry and enhancing the fishery. However, at the same time, the lake has also been used to process the effluent generated by upstream industry and waste water treatment works (WWTWs). Attempts to enhance the various uses of the lake are reviewed below in relation to their impact on the provision of other ecosystem services, with particular emphasis on conflicts of interest that have arisen amongst stakeholders. Such conflicts have often resulted from the fact that the lake is being used for both provisioning and regulating services at the same time, with the former being dependent upon good water quality and the latter causing a degradation of that water quality. In addition to the above, the economic costs and benefits of these management activities are also noted where sufficient data exist.

## Improving water supply to downstream users

### *Background*

Since the early 1700s, Loch Leven has supplied water to downstream industry for use either as a source of power or cooling, or in the manufacturing processes themselves (Munro, 1994). By the 1820s, 40 such industries, including corn mills, sawmills, paper mills, textile mills and bleaching fields, were using water supplied by the lake. As these industries expanded they began to find water in short supply, especially during the summer months. So, it was concluded that there was a need to manage the water from the lake more effectively to support these industries, and the jobs and income that they provided for the local community.

Various options were considered for improving water supply. Local industrialists preferred the option of raising the level of the lake and, thus, its storage capacity. In contrast, local landowners, whose land would be inundated if the level was increased, strongly supported the alternative option of lowering the level of the lake and building sluice gates to manage the rate of discharge in summer (Munro, 1994). It was argued that the latter option would have additional benefits, including the creation of new areas of highly productive farmland around the margins of the lake and the provision of a mechanism for controlling downstream flooding. The second option was approved by an Act of Parliament (1827).

### *Action*

Building and land drainage works began in 1831 to lower the level of the lake and install sluice gates on the outflow. By 1850, the water level had been lowered by about 1.5 m (Morgan, 1970), causing its surface area and mean depth to be reduced by 25 and 30%, respectively (Kirby, 1974). From this date onwards, the level of the lake was managed to ensure that the lake was full to capacity (i.e. water level ca. 107.3 m.a.o.d.) by late spring. Then, its discharge was controlled to ensure that the level of the lake fell only by about 0.18 m per month over the summer period (May & Carvalho, 2010). The total cost of these 'improvements' was about $US 60,000 at the time, a final figure that was more than double the original estimate (Munro, 1994). This is equivalent to a present day value of about $US 4.5M.

### *Consequences*

The drainage works successfully met their primary aim of providing a more stable and reliable water supply to downstream users (Sargent & Ledger, 1992). However, the secondary aim of increasing the area of land available for farming around the margins of the lake was only partially met. These works had been expected to provide an additional 440 ha. of good quality land (Committee for the Society for the Sons and Daughters of the Clergy, 1839), whereas, on completion, it was found that only 265 ha. of poor quality land had been reclaimed. In addition, the associated economic benefits were found to be 75% lower than expected (Munro, 1994).

The drainage works also affected a wide range of other ecosystem services that the lake provided. Firstly, amenity value was reduced because the local landowners claimed ownership of the recovered land and began to prosecute 'trespassers' found walking along the shore of the lake (Munro, 1994). Secondly, newly erected fencing on the reclaimed land prevented local villagers having access to peatlands that had long been used as a source of fuel, and to reed beds that had previously provided resources for the thatching of house roofs (Munro, 1994). Thirdly, it was estimated that the value of the fishery had been permanently reduced by about 33%, because the new shoreline was less suitable as a feeding ground for fish (Fleming, 1936), especially pike and brown trout (Winfield et al., 2011). Fourthly, and in contrast to expectation, management of the outflow proved to be more difficult than anticipated and actually increased the incidence of downstream flooding, at least initially. This led to several claims for compensation payments from disgruntled farmers whose land was flooded (Munro, 1994). Finally, the general ecology of the lake was disrupted, with macrophytes, such as *Isoetes* and *Chara* being lost (Salgado et al., 2009), Arctic charr becoming locally extinct (Burns-Begg, 1874), and the numbers of wading birds falling (Munro, 1994).

## Enhancing the fishery

### *Background*

Loch Leven has been an important fishery since 1314, when the Abbot of Dunfermline was granted permission to set fishing nets (Thorpe, 1974). In 1633, it was deemed necessary to protect the brown trout (*Salmo trutta* L.) stocks in the feeder streams and a private Act of Parliament, banning poaching, was introduced. By the early eighteenth century, fish stocks appeared to be thriving, with the lake being declared to be 'full of fish, particularly the finest trouts in the world' by Defoe (1723). Throughout the seventeenth and eighteenth centuries, fish were caught with nets and sold for local consumption (Thorpe, 1974). However, with the development of the railway in the nineteenth century, economically viable markets opened up further afield in large cities, such as Liverpool, Manchester and London (Munro, 1994) and fishing effort was increased. This increased pressure on the resource.

In contrast to the net fishery, which focused on the consumer market, rod fishing (angling) was introduced as a leisure activity in 1844. Enhanced by the arrival of the railway in the 1850s, the sports fishery had become an important generator of income for the owners by 1859. By the late 1800s, Loch Leven had become a world renowned recreational trout fishery (D'Arcy et al., 2006). In parallel to this development, the importance of net fishing declined, ceasing completely in 1873. Because of the local economic importance of the recreational fishery at Loch Leven, several attempts were made to enhance fish stocks and improve angling catches between the late 1800s and 2006.

### *Action*

A range of management activities were implemented to support the recreational fishery. These are documented in detail by Montgomery (1994) and summarised by Winfield et al. (2011). The main focus of these activities was fish stocking, with hatchery and rearing ponds coming into use in the mid to late 1800s. By 1882, an estimated 60,000 fry and 4,000 two-year-old brown trout per year, from breeding ponds at the nearby Howietoun fishery (Montgomery, 1994), were being placed in the lake's inflow streams. Although these trout were of "unrecorded origin", it seems likely that the original source of fish for this hatchery was Loch Leven (Day, 1887) because many of the ova produced were marketed as such and exported across the world. By the 1920s and early 1930s, up to 300,000 fry were being released each year. It seems likely that these, too, were reared from local stocks (Montgomery, 1994).

The hatchery closed in the late 1930s, but reopened in 1983 following a dramatic decline in fishery performance (Winfield et al., 2011). In that year, 5,000 fish of local origin were stocked directly into the lake (Montgomery, 1994). The number of stocked trout increased each year until 1988, when 166,000 were introduced. Levels of stocking remained in excess of 100,000 brown trout per year until 2004, when the number was reduced to 5,000. Stocking with brown trout was discontinued in 2006. In addition to stocking with brown trout, 30,000 non-native rainbow trout (*Oncorhynchus mykiss*) were introduced per year between 1993 and 2004 in an attempt to reduce the ongoing decline of the fishery and reflect changes in angler preferences (Winfield et al., 2011).

*Consequences*

It is difficult to assess the benefits of the stocking programme that was undertaken between 1882 and 1936 to the fishery, because there are few records of fish catches over that period. However, the reintroduction of fish stocking in the 1980s does seem to have been partially effective in that it appears to have slowed down the long-term decline in fish catches at the lake, for a few years at least (Winfield et al., 2011). Nevertheless, this apparent response was short-lived and catches began to decline further in the early 1990s.

The introduction of rainbow trout in the early 1990s was, initially, deemed a success from a fisheries point of view, because the catch of rainbow trout almost equalled that of brown trout in the first year of stocking and then greatly exceeded it in subsequent years (Winfield et al., 2011). However, by 2003, this figure had fallen dramatically and, with only about 10% of the stocked fish being caught by anglers, stocking with rainbow trout was suspended (Montgomery, 2004). By this time, the fishery was losing more than $US 150,000 per year and its operations were scaled down by about 75% (Loch Leven Fisheries, pers. comm.).

It was concluded that fish stocking had been ineffective in improving the economic value of the Loch Leven fishery in the longer term. Baer & Brinker (2010) came to a similar conclusion when studying the development of the brown trout fishery in the River Wutach, Germany, after fish stocking ceased in 2001. The results of these studies suggest that augmentation of a natural fishery with stocked individuals does not, necessarily, result in an improvement in the fishery or in angler satisfaction.

The reason for the failure of fish stocking to improve the fishery at Loch Leven is unclear. A range of theories have been suggested, but the one that has received most public attention, and caused most controversy, is the assertion that stocking the lake with small fish may have simply attracted more fish eating birds, primarily cormorants (*Phalacrocorax carbo*), to the site. While Wright (2003) concluded that cormorants had no effect on brown trout abundance or fishery performance in Loch Leven, Stewart et al. (2005) estimated that cormorants probably consumed about 80,800 brown and 5,200 rainbow trout over a 7-month period. These results suggested that there was considerable potential for competition between the birds and the fishery for available fish. Stewart et al. (2005) also found a relationship between the level of brown trout stocking in spring and the abundance of cormorants on the lake the following winter. This added weight to the argument that increased stocking may have attracted more cormorants to the lake, causing an increase in predation that negated any of the potential benefits that the stocking programme could have had on fishery yield. As there is considerable uncertainty about the size of the fish populations in Loch Leven, it is not possible to estimate the impact of piscivorous birds on the trout population as a whole (Winfield et al., 2011). However, the figures obtained by Stewart et al. (2005) indicated that almost 20% of the rainbow trout stocked into the lake by fisheries managers were being removed by cormorants. This equates to an economic cost of about $US 7,100 per year (Loch Leven Fisheries, pers. comm.).

The introduction of the rainbow trout coincided with an apparent change in the ecosystem functioning of the lake, as reflected by a significant increase in the chlorophyll *a*:total phosphorus (TP) ratio from the early 1990s onwards (Fig. 1). This may reflect changes in grazing pressure on algae from the zooplankter, *Daphnia*, which forms part of the diet of rainbow trout at this site (Duncan, 1994). Therefore, stocking with rainbow trout may have increased the likelihood of algal blooms developing in the lake at the same time as significant sums of money, i.e. about $US 7.1M, were being invested in reducing the likelihood of algal blooms by lowering the nutrient input to the lake from sources within the catchment (May et al., 2011).

Finally, stocking the lake with brown trout from the Howietoun rearing ponds in the late 1800s, or with imported rainbow trout between 1993 and 2004, may have affected the biodiversity of the resident fish population. However, the rainbow trout were unable to establish a viable population, and any other introduced trout were of the same species as those already in the lake. So, the fish stocking programme probably had less of an impact on the biodiversity of the system than some of the other management activities implemented, such as lowering the level of the lake and controlling the outflow (see above).

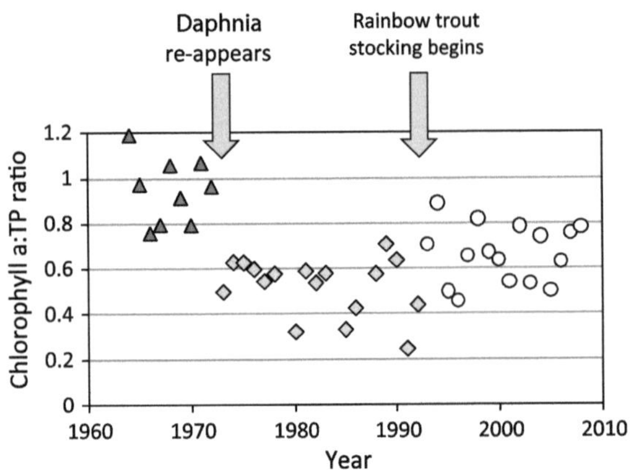


**Fig. 1** Chlorophyll *a*:total phosphorus (TP) ratio in Loch Leven showing changes that occurred when *Daphnia* re-appeared in the early 1970s and when rainbow trout were introduced in 1993. Differences in three treatment populations (i.e. 1964–1972: mean = 0.95, standard deviation (SD) = 0.15); 1973–1992: mean = 0.52, SD = 0.12; 1993–2008: mean = 0.66, SD = 0.13) were tested using analysis of variance (total degrees of freedom = 41; $F = 30.35$; $P < 0.001$). Each population was assumed to be normally distributed (i.e. $P > 0.05$) following an Anderson–Darling test. Tukey post-hoc analysis indicated significant differences between all populations (after May et al., 2007)

## Better management of waste from industry and WWTWs

### *Background*

In the early 1800s, the catchment of Loch Leven underwent a significant increase in population size. The local town of Kinross, for example, grew from about 2,100 inhabitants in 1801 to more than 3,200 inhabitants in 1851. This, together with an associated increase in local industries and more intensive farming activity within the catchment, led to more polluted effluent and runoff entering the lake via its inflows. At that time, there was an underlying assumption that the natural drainage system would simply carry the waste away, and little consideration was given to the likely impacts of these discharges on the lake and the ecosystem services that it provides.

The development of textile mills on one of the main inflows, the South Queich, has been a particular problem in terms of impacts on water quality and ecosystem services. One of these mills, which was established in 1867 and continues to the present day (Munro, 1994), discharged high levels of phosphorus (P) and pesticides into the lake for many years. The P-laden effluent resulted from manufacturing processes that used sodium hexametaphosphate, tetrasodium pyrophosphate and phosphoric acid (Holden & Caines, 1974), while the pesticide residues resulted from the use of the mothproofing agents Dielmoth (Dieldrin) from the 1950s to 1964, Eulan WANew (Chlorphenylid) from 1964 to 1980 and Mitin LP (Chorphenylid & Flucofenwon) from 1980 to 1988 (D'Arcy et al., 2006). Discharges of P from this source amounted to 2.6–7.8 t P year$^{-1}$ in the late 1960s/early 1970s (Holden & Caines, 1974) and about 6.3 t P year$^{-1}$ by the mid 1980s (Bailey-Watts & Kirika, 1987), i.e. about 50–60% of the annual P input to the lake in the earlier years and about 30% of the input in the later years. The exact level of discharge of pesticide residues is unknown.

Treated effluent from WWTWs within the catchment was also a significant source of P input to the lake. In the late 1960s/early 1970s, this contribution was estimated to be about 1.7 t P year$^{-1}$, or 20% of the total P input to the lake (Holden & Caines, 1974). By 1985, this figure had risen to 5.3 t P year$^{-1}$, or 27% of the total input (Bailey-Watts & Kirika, 1987). Other sewage-related sources, such as septic tanks, have also been estimated to contribute 10–14% of the overall P input to this system (Frost, 1996; Dudley & May, 2007).

Using the lake and its inflows for the disposal of both nutrient laden waste and pesticide discharges had a marked detrimental effect on the lake's ecological structure The increased supply of P resulted in excessive algal growth (Bailey-Watts & Kirika, 1987), while pesticide residues (especially from dieldrin) appear to have reduced the ability of *Daphnia* to control algal biomass through grazing (Holden, 1966; D'Arcy et al., 2006). Although the evidence for the latter is mainly circumstantial, it is clear that *Daphnia*, which had been common in the lake before the 1950s (Scott, 1891, 1899; Morgan, 1970), disappeared from the system at about the same time as the mill began discharging dieldrin. It reappeared in the early 1970s (Leven IBP Project Report, 1970–1971), when these discharges had stopped and levels in the lake, as indicated by fish tissue analyses, had dropped by about 99% (Wells & Cowan, 1984).

Together, the two pollutants described above appear to have encouraged troublesome algal blooms to develop in the lake by increasing algal productivity and reducing grazing losses at the same time. These blooms decreased water clarity, lowering the aesthetic, economic and amenity values of the lake, and reducing the depth to which rooted, submerged vegetation was able to grow (May & Carvalho, 2010; Dudley et al., 2011). The blooms also limited the ability of the lake to support other ecosystem services, such as supplying clean water and providing good quality habitat for aquatic birds, especially those that are dependent on underwater plants (Allison & Newton, 1974). By the early 1980s, it had become clear that management intervention was required to improve the water quality of the lake and the wide range of ecosystem services that had been damaged by both nutrient pollution and pesticide residues. Although the economic cost of the earlier water quality problems have not been documented, it has been estimated that a single algal bloom in 1992 cost the local community more than $US 1.5M in lost revenue over a single summer period.

*Action*

Although several different mothproofing agents had been used by the woollen mill and discharged into the lake over many years, the main ecological impacts seem to have been related to the use/discharge of Dielmoth (dieldrin) from the early 1950s until 1964. This product was initially replaced by other mothproofing agents (see above) but, in 1988, the mill voluntarily stopped using mothproofers altogether (D'Arcy et al., 2006).

Phosphorus inputs to the lake were reduced significantly between 1985 (ca. 20 t year$^{-1}$) and 1995 (ca. 8 t year$^{-1}$) (Bailey-Watts & Kirika, 1999), mainly by reducing outputs from the mill and the WWTWs. The former was achieved initially by effluent diversion and then by changing the chemical processes used; the latter was achieved by upgrading the works and introducing tertiary treatment (i.e. P-stripping). The total cost of these improvements was estimated to be more than $US 6.5M (LLCMP, 1999). In addition to controlling these larger point source discharges of P, households in rural areas of the catchment were also encouraged to manage their septic tanks more effectively, and more stringent planning regulations were put into place to reduce the likelihood of P discharges from new housing developments in unsewered areas polluting the lake (LLCMP, 1999; Dudley & May, 2007).

*Consequences*

Reducing the discharge of dieldrin based pesticides to the lake had a significant effect on ecosystem functioning. *Daphnia*, the main grazer of algae in the lake and an important source of food for fish, re-appeared in the summer of 1970 and, by July 1971, had achieved a maximum population density of 74.4 ind. l$^{-1}$ (Johnson & Walker, 1974). Although there is no direct evidence of this link, *Daphnia* are known to be particularly sensitive to some pesticides because they accumulate larger amounts of these substances in their body tissue than other aquatic invertebrates (Walsh, 1978). *Daphnia magna*, for example, concentrated photodieldrin by a factor of x 63,000 when exposed to contaminated media by Khan & Khan (1974) and their reproductive ability was inhibited when they were exposed to other pesticides, such as DDT, by Maki & Johnson (1975). Sprules (1975) has shown that zooplankton species vary in their sensitivity to pollutants and that this can affect community structure. It seems likely that dieldrin pollution had a similarly differential effect on zooplankton species composition in Loch Leven, reducing the number of *Daphnia* significantly while apparently having little impact on other species, such as *Cyclops* (Johnson & Walker, 1974). Even though the discharge of dieldrin stopped in 1964, low, but discernable, concentrations of up to 26 μg kg$^{-1}$ were still being detected within the crustacean zooplankton community up to 15 years later (Wells & Cowan, 1984).

In addition to the potential impact of pesticides on *Daphnia*, fish from the lake were found to have relatively high dieldrin concentrations in both their muscle and liver tissues during the 1960s (Holden, 1972). In 1964, those levels amounted to 0.34 and 0.3 mg kg$^{-1}$, respectively, for trout, 0.78 and 1.2 mg kg$^{-1}$, respectively, for perch, and 0.38 and 6.1 mg kg$^{-1}$, respectively, for pike (Wells & Cowan, 1984). Following the change from dieldrin to Eulan WANew in 1964, fish tissue samples analysed between 1963 and 1970 showed a steady decline in dieldrin concentrations, with levels of residue having fallen by 99% after 4–7 years (Wells & Cowan, 1984). Although the biological effect of dieldrin on the ecology of the fish populations is unclear, the fact that they were known to be contaminated probably reduced the value of the fishery amidst concerns about the health risks to anglers of eating contaminated catches (D'Arcy et al., 2006).

As a result of this sudden increase in *Daphnia* abundance in the early 1970s, grazing losses increased and chlorophyll *a* concentrations fell (Bailey-Watts, 1974). This led to a significant lowering of the chlorophyll *a*:TP ratio in the lake (Fig. 1) and an associated reduction in the frequency of algal blooms. This caused a temporary improvement in water clarity and an increase in the abundance of submerged macrophytes (Johnson & Walker, 1974; Dudley et al., 2011). However, as P inputs to the lake continued to increase (May et al., 2011), algal growth outstripped loss processes, even with the increase in zooplankton grazing, and algal blooms became more common, again.

Amidst concerns about the continuing decline in water quality at the lake, and its impacts on the fishery, tourism and downstream industry, targets were set for the restoration of this waterbody in 1993. These were based on P and chlorophyll *a* concentrations, water clarity and macrophyte growing depths (D'Arcy et al., 2006; Carvalho et al., 2011). However, when P discharges from WWTWs and the mill were reduced by 60% between 1993 and 1999, the expected recovery in water quality did not occur immediately. Instead, a prolonged recovery trajectory was observed as a result of internal release of phosphorus from the lake sediments (Spears et al., 2011). During the recovery phase, open water P concentrations have fallen (Carvalho et al., 2011), water clarity has increased (especially in spring), and submerged macrophytes have been observed to recolonise deeper areas of the lake (May & Carvalho, 2010). These improvements were associated with increases in species composition and abundance of macrophytes and invertebrates (Dudley et al., 2011; Gunn et al., 2011) and an improvement in the habitat available to fish communities (Winfield et al., 2011).

### *Conflicts and trade-offs*

Throughout the documented history of management intervention at Loch Leven, it is clear that changes focused on improving one particular ecosystem service have been made with little or no prior consideration of the knock-on effects on others. In some cases, these effects have been significant. The most important of these are summarised below.

The impact of the drainage works on ecosystem services at Loch Leven is summarised in Fig. 2. While

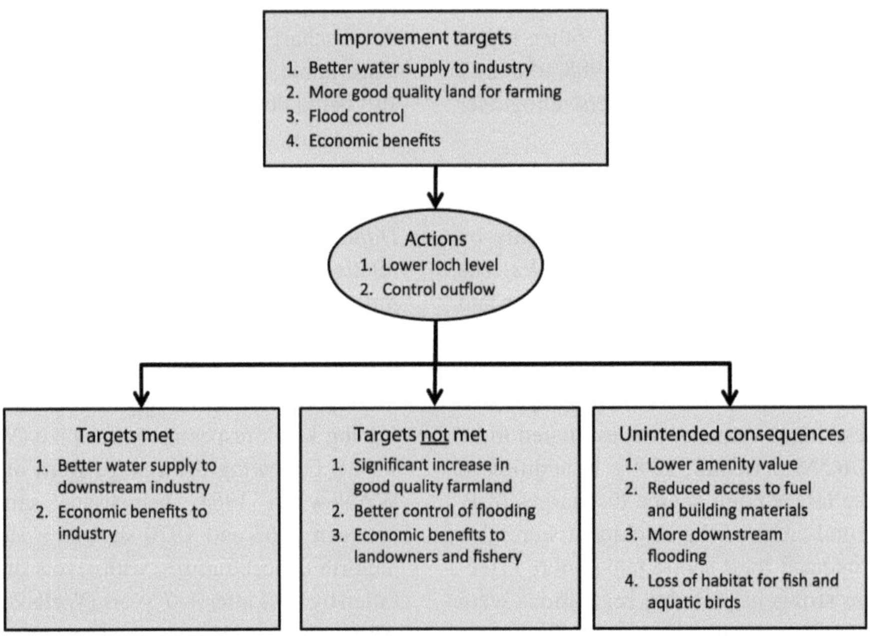


**Fig. 2** Summary of the impacts of water level change and outflow management on ecosystem services at Loch Leven

these works improved the water supply to downstream industry, resulting in economic benefits to mill owners as intended, they did not deliver the expected increase in good quality farmland, better control of downstream flooding or economic benefits to local landowners. Instead, changes in the water level of the lake led to increased downstream flooding, the loss of marginal macrophyte beds, damage to fish habitat and food supply, and lower amenity value. In addition, fencing of reclaimed areas of land restricted the access of local people to traditional sources of fuel and building materials.

Fish stocking aimed to boost the fishery in terms of increasing angling catches and improving its economic value. However, there is little evidence that either of these aims was achieved successfully and it is possible that this activity may have led to detrimental effects on other ecosystem services provided by the lake. These include reducing the ability of *Daphnia* to control algal blooms through their grazing activity, lowering the aesthetic and economic value of the lake by attracting large numbers of cormorants, and undermining the conservation value of the lake through the introduction of non-native species. These effects are summarised in Fig. 3.

Using Loch Leven and its tributaries to dispose of nutrient and pesticide wastes from industry and WWTWs also affected the ecosystem services that the lake provides. When these inputs were reduced, P concentrations decreased, water clarity improved (at least in spring) due to the re-appearance of key zooplankton grazers, and macrophytes began to grow in deeper water. In addition, the level of pesticide residues within the zooplankton and fish communities fell, more extensive macrophyte beds developed providing a better habit for fish and their invertebrate food, and the biodiversity of many plant and animal communities increased. These responses are summarised in Fig. 4.

The results outlined above, and summarised in Table 1, clearly demonstrate that management for the benefit of one ecosystem service can often be at the expense of a whole range of others, as suggested by Holling & Meffe (1996). So, it is important to understand the relationships that exist across all of the services that an ecosystem provides before deciding on an appropriate management strategy. Balancing these effects, or considering ecosystem service 'trade-offs', must take into account the type, magnitude, and relative mix of services that ecosystems can provide

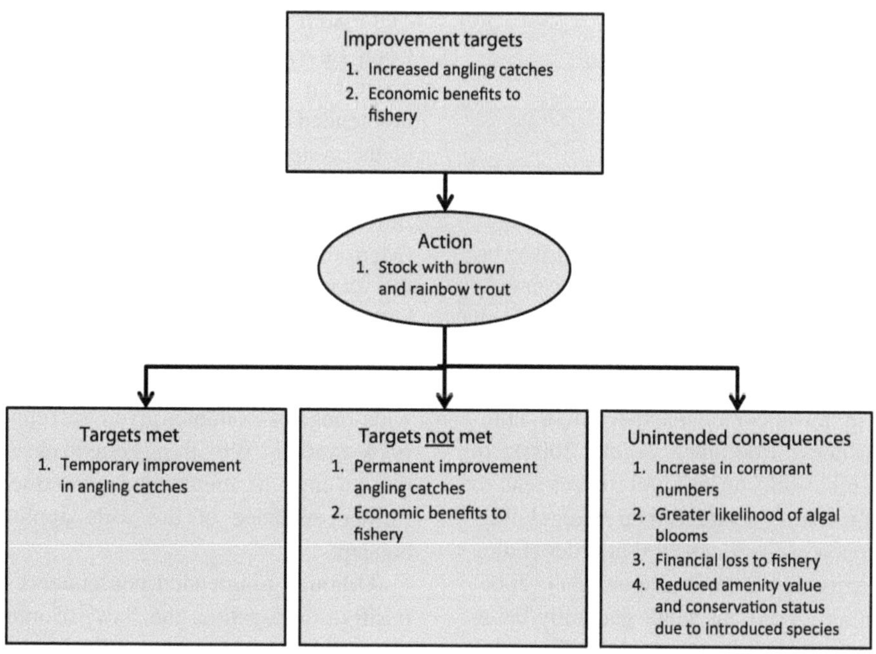


**Fig. 3** Summary of the impacts of fish stocking on ecosystem services at Loch Leven

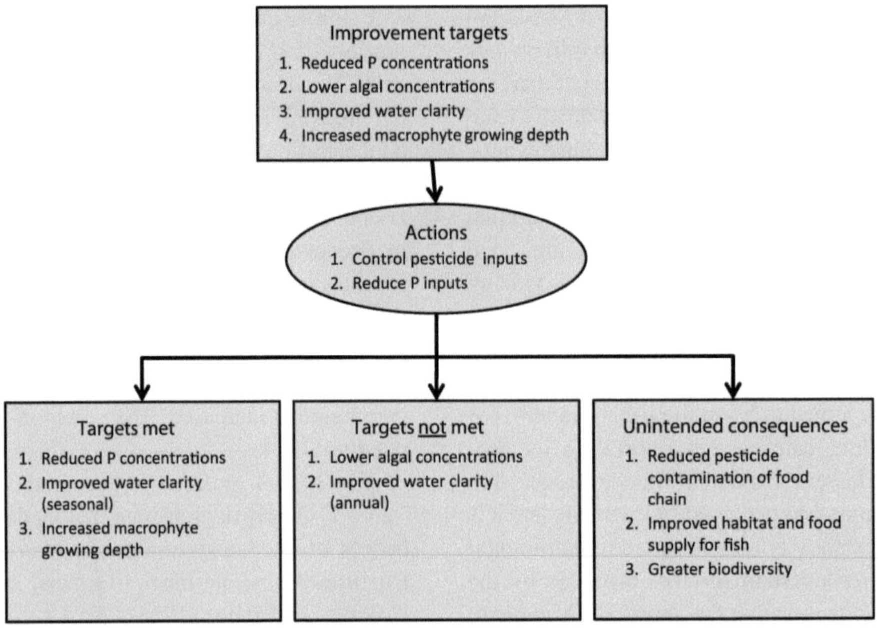


**Fig. 4** Summary of the impacts of reductions in pesticide and phosphorus inputs on ecosystem services at Loch Leven

**Table 1** Summary of the positive (+), negative (−) and neutral (±) effects of management for the benefit of one key ecosystem service (Column 1) at Loch Leven on the provision of other such services (Row 1)

| | Water supply | Waste management | Flood control | Recreation |
|---|---|---|---|---|
| Water supply | + | − | − | − |
| Waste management | − | + | ± | − |
| Flood control | − | − | + | − |
| Recreation | ± | − | − | + |

under different management scenarios. However, realistically, such informed decision making can only be applied effectively to systems that are well understood; if not, other 'trade-offs' may arise without premeditation or even awareness that they are taking place, leading to unwelcome, unexpected or unintended consequences (Rodríguez et al., 2006). In managing these trade-offs, human preferences tend to focus first on provisioning services, then on regulating, cultural, and supporting services, in that order (Foley et al., 2005; Pereira et al., 2005; Rodríguez et al., 2006; van Jaarsveld et al., 2005), the latter generally being "taken for granted" (Rodríguez et al., 2006). This approach is unlikely to provide sustainable solutions to the management of ecosystem services, as this study on Loch Leven demonstrates.

## Discussion

The results from the long-term monitoring studies at Loch Leven clearly show that the management of ecosystem services often results in unexpected or unintended consequences, especially if the complexity of the system being managed is not sufficiently well understood. Such outcomes show that the so-called 'law' of unintended or unanticipated consequences (Merton, 1936; Norton, 2010), which is widely used in the social sciences (e.g. economics, history, philosophy, political science and sociology) but less commonly applied to the environmental sciences, is equally applicable to the management of lakes. The wide range of examples given by Tenner (1997) in his book entitled "Why things bite back: Technology and the revenge of unintended consequences", provide further evidence of the wide applicability of this concept.

Although unintended consequences can be either positive or negative, the 'law' of unintended consequences warns that intervention in a complex system without taking into consideration the relationships between one component and another always creates unanticipated, and often undesirable, results. Merton (1936) suggests that there are five possible causes of unexpected consequences when management actions are planned and implemented. These are as follows:

- Lack of knowledge, leading to incomplete analysis of the problem
- Error, leading to incorrect analysis of the problem
- Immediate interest, which may override long-term interests
- Basic values, which may prohibit certain actions even if the long-term result might be unfavourable
- Self-defeating prophecy, whereby fear of the consequences drive people to find solutions before the problem occurs

The examples from Loch Leven illustrate how the management of ecosystem services in the natural environment can lead to a wide range of unintended consequences. In particular, they show how Causes 1–3, i.e. lack of knowledge, incorrect analysis of the problem and addressing immediate (especially economic) interests, can cause a wide range of problems that threaten the long-term sustainability of the system and the services that it provides. In contrast, Causes 4 and 5 have been less applicable to problems that have arisen from management intervention at this site over the years.

When unintended consequences occur, it is often difficult to turn back the clock. At Loch Leven, for example, the drainage works and consequent change in water level caused a wide range of ecological and hydrological problems, but it is unlikely that this significant amount of engineering work will ever be reversed. Even if it were, it is doubtful that this would result in the complete restoration of the lake's ecology to pre-drainage conditions without considerable management intervention. This issue of reversibility, which is very important when considering restoration options for impacted ecosystems, has also been raised in relation to the management of many other systems, such as the River Rhine. Huisman (1995) showed that the concept of reversibility, i.e. the "clean up the water and life will return" approach, did not work when waste water treatment was improved and levels of pollution in the river were reduced. He concluded that damaged ecosystems and the services that they provide could only be restored through a significant change of approach, i.e. from a one-sided promotion of individual interests to a more integrated and sustainable method of management.

It is now generally accepted that the ecological structure and function of lakes will be altered as a result of anthropogenic climate change (Carpenter et al., 1992). In Loch Leven, climate induced changes in ecological structure are expected to be driven mainly by wind speed and direction (Spears & Jones, 2010), temperature (Carvalho et al., 2011) and rainfall (Carvalho et al., 2011). Additionally, the performance of pivotal ecosystem functions, such as internal nutrient cycling (Spears et al., 2008; Spears et al., 2011), zooplankton seasonality and abundance, and, therefore, grazing pressure (Ferguson et al., 2008), are expected to vary in Loch Leven as a result of climate change. However, the importance of other key drivers of biodiversity, including land use change, atmospheric deposition and biotic exchange (which is expected to be the most important driver of global freshwater biodiversity loss in this century according to Sala et al., 2000), are not well understood in Loch Leven. It is clear that in order to mitigate against decline, and ideally to maximise the future provision of ecosystem services in Loch Leven, a better understanding of the drivers of feedbacks between biodiversity and ecosystem functioning is required (Covich et al., 2004).

## Conclusion

More than 40 years ago, Ehrlich & Ehrlich (1970) correctly warned mankind that "the most subtle and dangerous threat to man's existence … is the potential destruction, by man's own activities, of those ecological systems upon which … the human species depends". For this reason, when management changes aimed at enhancing one particular ecosystem service are being considered, it is important that potentially damaging effects on others are also taken into account. In practice, taking this approach requires a better understanding of the role and complexity of ecosystem function, and of species interactions in delivering ecosystem services, than is currently available. While further research is clearly needed, the value of long-term datasets in providing knowledge and understanding through 'hindsight' should not be underestimated. Valuable insight can be gained from studies such as that on Loch Leven to help ensure that future decision making takes into consideration the full range of benefits expected and the likelihood of unintended, especially harmful, effects on the delivery of other ecosystem services. Our results support the view of Rodríguez et al. (2006) that, in the longer term, successful management policies are likely to be those that incorporate lessons learned from previous decisions into future management actions.

**Acknowledgments** We thank Kinross Estates and the Loch Leven Fisheries for their help with this work and for providing access to Loch Leven and fisheries records. This research was funded by Natural Environment Research Council, UK with some small contributions from Scottish Natural Heritage and the Scottish Environment Protection Agency. Loch Leven is part of the UK Environmental Change Network (http://www.ecn.ac.uk/).

## References

Act of Parliament, 1827. The River Leven Improvement Act.

Allison, A. & I. Newton, 1974. Waterfowl at Loch Leven, Kinross. Proceedings of the Royal Society of Edinburgh, B 74: 365–381.

Baer, J. & A. Brinker, 2010. The response of a brown trout stocks and perception of anglers to cessation of brown trout stocking. Fisheries Management and Ecology 17: 157–164.

Bailey-Watts, A. E., 1974. The algal plankton of Loch Leven, Kinross. Proceedings of the Royal Society Edinburgh, B 74: 136–156.

Bailey-Watts, A. E. & A. Kirika, 1987. A re-assessment of the phosphorus inputs to Loch Leven (Kinross, Scotland): rationale and an overview of results on instantaneous loadings with special reference to runoff. Transactions of the Royal Society of Edinburgh, Earth Sciences 78: 351–367.

Bailey-Watts, A. E. & A. Kirika, 1999. Poor water quality in Loch Leven (Scotland) in 1995, in spite of reduced phosphorus loadings since 1985: the influences of catchment management and inter-annual weather variation. Hydrobiologia 403: 135–151.

Baron, J. S., N. L. Poff, P. L. Angermeier, C. N. Dahm, P. H. Gleick, N. G. Hairston, R. B. Jackson, C. A. Johnston, B. D. Richter & A. D. Steinman, 2002. Meeting ecological and societal needs for freshwater. Ecological Applications 12: 1247–1260.

Burns-Begg, R., 1874. The Loch Leven Angler. Barnett, Kinross.

Carpenter, S. R., S. G. Fisher, N. B. Grimm & J. F. Kitchell, 1992. Global change and freshwater ecosystems. Annual Review of Ecology and Systematics 23: 119–139.

Carvalho, L., C. Ferguson, I. Gunn, H. Bennion, B. Spears, A. Kirika & L. May, 2011. Water quality of Loch Leven: responses to enrichment, restoration and climate change. Hydrobiologia. doi:10.1007/s10750-011-0923-x.

Castle, K., C. A. Frost & D. F. Flint, 1999. The Loch Leven Project–Buffer strips in practice on a catchment scale. Aspects of Applied Biology 54: 71–78.

Committee for the Society for the Sons and Daughters of the Clergy, 1839. The New Statistical Account of Scotland. Fife, Kinross.

Covich, A., M. C. Austen, F. Barlocher, E. Chauvet, B. J. Cardinale, C. L. Biles, P. Inchausti, O. Dangles, M. Solan, M. O. Gessner, B. Statzner & B. Moss, 2004. The role of biodiversity in the functioning of freshwater and marine benthic ecosystems. BioScience 54: 767–775.

D'Arcy, B. J., L. May, J. Long, I. R. Fozzard, S. Greig & A. Brachet, 2006. The restoration of Loch Leven, Scotland, UK. Water Science Technology 53: 183–191.

Day, F., 1887. On the Loch-Leven trout (*Salmo levenensis*). Journal of the Linnaean Society of London, Zoology 20: 71–90.

Defoe, D., 1723. A Journey Through Scotland. London.

Dudley, B. & L. May, 2007. Estimating the phosphorus load to waterbodies from septic tanks. Report to the Scottish Environment Protection Agency and Scottish Natural Heritage: 45 pp.

Dudley, B., I. D. M. Gunn, L. Carvalho, I. Proctor, M. T. O'Hare, K. J. Murphy & A. Milligan, 2011. Changes in aquatic macrophyte communities in Loch Leven—evidence of recovery from eutrophication? Hydrobiologia. doi:10.1007/s10750-011-0924-9.

Duncan, W., 1994. The fish community in Loch Leven. In Hutchinson, P. & A. Walker (eds), The Loch Leven Trout Fishery—Its Future. Institute of Fisheries Management, Pitlochry: 29–38.

Ehrlich, P. R. & A. Ehrlich, 1970. Population, Resources, Environment: Issues on Human Ecology. W.H. Freeman, San Francisco.

Ferguson, C. A., L. Carvalho, E. M. Scott, A. W. Bowman & A. Kirika, 2008. Assessing ecological responses to environmental change using statistical models. Journal of Applied Ecology 45: 193–203.

Fleming, J., 1936. Interim report by the Reverend Dr Fleming as valuator of the damage done to the fishings by the drainage of Loch Leven, Kinross. Manuscript, Sheriff Court, Kinross.

Foley, J. A., R. DeFries, G. P. Asner, C. Barford, G. Bonan, S. R. Carpenter, F. S. Chapin, M. T. Coe, G. C. Daily, H. K. Gibbs, J. H. Helkowski, T. Holloway, E. A. Howard, C. J. Kucharik, C. Monfreda, J. A. Patz, I. C. Prentice, N. Ramankutty & P. K. Snyder, 2005. Global consequences of land use. Science 309: 570–574.

Folke, C., S. Carpenter, B. Walker, M. Scheffer, T. Elmqvist, L. Gunderson & C. S. Holling, 2004. Regime shifts, resilience, and biodiversity in ecosystem management. Annual Review of Ecology Evolution and Systematics 35: 557–581.

Frost, A., 1996. Loch Leven and diffuse pollution. In Petchey, A., B. D'Arcy & C. Frost (eds), Diffuse Pollution and Agriculture. Scottish Agricultural College, Aberdeen: 174–182.

Gunn, I. D. M., M. T. O'Hare, P. S. Maitland & L. May, 2011. Long-term trends in Loch Leven invertebrate communities. Hydrobiologia. doi:10.1007/s10750-011-0926-7.

Heal, G., G. C. Daily, P. R. Ehrlich, J. Salzman, C. Boggs, J. Hellman, J. Hughes, C. Kremen & T. Ricketts, 2001. Protecting natural capital through ecosystem service districts. Stanford Environmental Law Journal 20: 333–364.

Holden, A. V., 1966. Organochlorine insecticide residues in salmonid fish. Journal of Applied Ecology 3(Suppl): 45–53.

Holden, A. V., 1972. The effects of pesticides on life in freshwater. Proceedings of the Royal Society London, Series B 180: 383–394.

Holden, A. V. & L. A. Caines, 1974. Nutrient chemistry of Loch Leven. Proceedings of the Royal Society Edinburgh (B) 74: 101–122.

Holling, C. S. & G. K. Meffe, 1996. Command and control and the pathology of natural resource management. Conservation Biology 10: 328–337.

Huisman, P., 1995. From one-sided promotion of individual interests to integrated water management in the Rhine basin. Water Science and Technology 31: 59–66.

Johnson, D. & A. F. Walker, 1974. The zooplankton of Loch Leven, Kinross. Proceedings of the Royal Society of Edinburgh (B) 74: 285–294.

Khan, H. M. & M. A. Q. Khan, 1974. Biological magnification of photodieldrin by food chain organisms. Archives of Environmental Contamination and Toxicology 2: 289–301.

Kirby, R. P., 1971. The bathymetrical resurvey of Loch Leven, Kinross. Geography Journal 137: 372–378.

Kirby, R. P., 1974. The morphological history of Loch Leven, Kinross. Proceedings of the Royal Society Edinburgh, B 74: 57–68.

LLCMP, 1999. Loch Leven Catchment Management Plan: The Report of the Loch Leven Area Management Advisory Group: 93 pp.

Loch Leven IBP Project Report, 1970–1971. Unpublished report: 22 pp.

Maki, A. W. & H. E. Johnson, 1975. Effects of PCB ('Aroclor' 1254) and p,p′-DDT on production and survival of Daphnia magna Strauss. Bulletin of Environmental Contamination and Toxicology 13: 412–416.

May, L. & B. M. Spears, 2011. A history of scientific research at Loch Leven, Kinross, Scotland. Hydrobiologia. doi: 10.1007/s10750-011-0929-4.

May, L. & L. Carvalho, 2010. Maximum growing depth of macrophytes in Loch Leven, Scotland, United Kingdom, in relation to historical changes in estimated phosphorus loading. Hydrobiologia 646: 123–131.

May, L., L. Carvalho, I. D. M. Gunn & A. Kirika, 2007. The impact of trophic interactions on the recovery of Loch Leven after reduction in phosphorus loads. In Heckrath, G., G. H. Rubaek & B. Kronvang (eds), Diffuse Phosphorus Loss. Risk assessment, Mitigation Options and Ecological Effects in River Basins. Aarhus, Denmark, National Environment Research Institute, University of Aarhus: 71–73, 4 pp. (DJF Plant Science, 130).

May, L., L. H. Defew, H. Bennion & A. Kirika, 2011. Historical changes (1905–2005) in external phosphorus loads to Loch Leven, Scotland, UK. doi:10.1007/s10750-011-0922-y

Merton, R. K., 1936. The unanticipated consequences of purposive social action. American Sociological Review 1: 894–904.

Millennium Ecosystem Assessment (MEA), 2005. Ecosystems and Human Well-Being: Wetlands and Water Synthesis. World Resources Institute, Washington, DC.

Montgomery, D., 1994. Management of the Loch Leven trout fishery. In Hutchinson, P. & A. Walker (eds), The Loch Leven Trout Fishery—Its Future. Institute of Fisheries Management, Pitlochry: 26–28.

Montgomery, J., 2004. Loch Leven Fisheries—Executive Summary: 7 pp.

Morgan, N. C., 1970. Changes in the fauna and flora of a nutrient enriched lake. Hydrobiologia 35: 545–553.

Munro, D., 1994. Loch Leven and the River Leven—A Landscape Transformed. The River Leven Trust, Markinch: 196 pp.

Norton, R., 2010. Unintended Consequences. The Concise Encyclopedia of Economics, Library of Economics and Liberty. http://www.econlib.org/library/Enc/Unintended Consequences.html

Pereira, H. M., B. Reyers, M. Watanabe, E. Bohensky, S. Foale, C. Palm, M. V. Espaldon, D. Armenteras, M. Tapia, A. Rincón, M. J. Lee, A. Patwardhan & I. Gomes, 2005. Condition and trends of ecosystem services and biodiversity. In Capistrano, D., C. Samper, M. J. Lee & C. Raudsepp-Hearne (eds), Ecosystems and Human Well-being: Multi scale Assessments, Volume 4. Findings of the Sub-global Assessments Working Group of the Millennium Ecosystem Assessment. Island Press, Washington, DC.

Perth & Kinross Council, 2004. Kinross Area Local Plan 2004: 106 pp.

Reid W. V., H. A. Mooney, A. Cropper, D. Capistrano, S. R. Carpenter, K. Chopra, P. Dasgupta, T. Dietz, A. K. Duraiappah, R. Hassan, R. Kasperson, R. Leemans, R. M. Rodríguez, J. P., T. D. Beard Jr., J. Agard, E. Bennett, S. Cork, G. Cumming, D. Deane, A. P. Dobson, D. M. Lodge, M. Mutale, G. C. Nelson, G. D. Peterson & T. Rodríguez, J. P., T. D. Beard Jr., E. M. Bennett, G. S. Cumming, S. Cork, J. Agard, A. P. Dobson & G. D. Peterson, 2006. Trade-offs across space, time, and ecosystem services. Ecology and Society 11: 28. http://www.ecologyandsociety.org/vol11/iss1/art28/.

Rodríguez, J. P., T. D. Beard, Jr., E. M. Bennett, G. S. Cumming, S. Cork, J. Agard, A. P. Dobson, and G.D. Peterson. 2006. Trade-offs across space, time, and ecosystem services. Ecology and Society 11: 28, 14pp. http://www.ecologyand society.org/vol11/iss1/art28/

Sala, O. E., F. S. Chapin, J. J. Armesto, E. Berlow, J. Bloomfield, R. Dirzo, E. Huber-Sanwald, L. F. Huenneke, R. B. Jackson, A. Kinzig, R. Leemans, D. M. Lodge, H. A. Mooney, M. Oesterheld, N. L. Poff, M. T. Sykes, B. H. Walker, M. Walker & D. H. Wall, 2000. Global biodiversity scenarios for the year 2100. Science 287: 1770–1774.

Salgado, J., S. Sayer, L. Carvalho, T. Davidson & I. Gunn, 2009. Assessing aquatic macrophyte community change through the integration of palaeolimnological and historical data at Loch Leven, Scotland. Journal of Paleolimnology 43: 191–204.

Sargent, R. J. & D. C. Ledger, 1992. Derivation of a 130 year runoff record for the Loch Leven catchment, south-east Scotland. Proceedings of the Institute of Civil Engineers Water, Maritime and Energy 96: 71–80.

Scott, T., 1891. The invertebrate fauna of the inland waters of Scotland. Part II. In 9th Annual Report of the Fisheries Board of Scotland, Part III: 269–296.

Scott, T., 1899. The invertebrate fauna of the inland waters of Scotland—report on special investigation. In 17th Annual Report of the Fisheries Board of Scotland, Part III: 161–165.

Spears, B. M. & I. Jones, 2010. The long-term (1979–2005) effects of the north Atlantic oscillation on wind induced wave mixing in Loch Leven (Scotland). Hydrobiologia 646: 49–59.

Spears, B. M., L. Carvalho, R. Perkins & D. M. Paterson, 2008. The effects of light on sediment nutrient flux and water column nutrient stoichiometry in a shallow lake. Water Research 42: 977–986.

Spears, B. M., L. Carvalho, R. Perkins, A. Kirika & D. M. Paterson, 2011. Long-term variation and regulation of internal loading in Loch Leven. Hydrobiologia. doi: 10.1007/s10750-011-0921-z

Sprules, W. G., 1975. Midsummer crustacean zooplankton communities in acid-stressed lakes. Journal of the Fisheries Research Board of Canada 32: 389–395.

Stewart, D. C., S. J. Middlemas, W. R. Gardiner, S. Mackay & J. D. Armstrong, 2005. Diet and prey selection of cormorants (*Phalacrocorax carbo*) at Loch Leven, a major stocked trout fishery. Journal of the Zoological Society of London 267: 191–201.

Tenner, E., 1997. Why things bite back: technology and the revenge of unintended consequences. Vantage Books, New York.

Thorpe, J. E., 1974. Trout and perch populations at Loch Leven, Kinross. Proceedings of the Royal Society of Edinburgh, B 74: 296–313.

van Jaarsveld, A. S., R. Biggs, R. J. Scholes, E. Bohensky, B. Reyers, T. Lynam, C. Musvoto & C. Fabricius, 2005. Measuring conditions and trends in ecosystem services at multiple scales: the Southern African Millennium Ecosystem Assessment (SAfMA) experience. Philosophical Transactions of the Royal Society B-Biological Sciences 360: 425–441.

Walsh, G. E., 1978. Chapter 12: Toxic effects of pollutants on plankton. In Butler, G. C. (ed.), Principles of Ecotoxicology. John Wiley & Sons, Inc., New York: 257–274.

Wells, D. E. & A. A. Cowan, 1984. Fate and Distribution of the Mothproofing Agents Dieldrin and Eulan WA New in Loch Leven, Kinross, 1964–1979. Environmental Pollution (Series B) 7: 11–33.

Winfield, I. J., Adams, C. E., Armstrong, J. D., Gardiner, R., Kirika, A., Montgomery, J., Spears, B. M., Stewart, D.C., Thorpe, J. E. & W. Wilson, 2011. Changes in the fish community of Loch Leven: untangling anthropogenic pressures. Hydrobiologia. doi:10.1007/s10750-011-0925-8.

Wright, G. A., 2003. Impact of cormorants on the Loch Leven trout fishery and the effectiveness of shooting as mitigation. In Cowx, I. G. (ed.), Interactions between fish and birds: implications for management. Fishing News Books. Blackwell Scientific Publications, Oxford: 288–297.

MIX
Papier aus verantwortungsvollen Quellen
Paper from responsible sources
FSC® C105338

If you have any concerns about our products,
you can contact us on
**ProductSafety@springernature.com**

In case Publisher is established outside the EU,
the EU authorized representative is:
**Springer Nature Customer Service Center GmbH**
**Europaplatz 3, 69115 Heidelberg, Germany**

Printed by Libri Plureos GmbH
in Hamburg, Germany